The Wounds That Heal

Heroism and Human Development

Judith A. Schwartz
and
Richard B. Schwartz

UNIVERSITY PRESS OF AMERICA,® INC.
Lanham • Boulder • New York • Toronto • Plymouth, UK

University Press of America,® Inc.
4501 Forbes Boulevard
Suite 200
Lanham, Maryland 20706
UPA Acquisitions Department (301) 459-3366

Estover Road
Plymouth PL6 7PY
United Kingdom

Printed in the United States of America
British Library Cataloging in Publication Information Available

Library of Congress Control Number: 2010927731
ISBN: 978-0-7618-5179-0 (paperback : alk. paper)
eISBN: 978-0-7618-5180-6

Cover photos, top to bottom: Jane Austen, George S. Patton, Jr., Winston Churchill, and T. E. Lawrence

∞™ The paper used in this publication meets the minimum requirements of American National Standard for Information Sciences—Permanence of Paper for Printed Library Materials, ANSI/NISO Z39.48-1992.

For our Fathers and our Son, Heroes First and Last

And it may not be amiss for all people who shall meet with my story to make this just observation from it—namely, how frequently in the course of our lives the evil which in itself we seek most to shun, and which, when we are fallen into it, is the most dreadful to us, is oftentimes the very means or door of our deliverance, by which alone we can be raised again from the affliction we are fallen into.

Daniel Defoe, *Robinson Crusoe*

There is a wisdom that is woe; but there is a woe that is madness. And there is a Catskill eagle in some souls that can alike dive down into the blackest gorges, and soar out of them again and become invisible in the sunny spaces. And even if he for ever flies within the gorge, that gorge is in the mountains; so that even in his lowest swoop the mountain eagle is still higher than other birds upon the plain, even though they soar.

Herman Melville, *Moby Dick*, xcvi

Contents

List of Figures

Preface

> The world has no room for cowards. We must all be ready somehow to toil, to suffer, to die. And yours is not the less noble because no drum beats before you when you go out into your daily battlefields, and no crowds shout about your coming when you return from your daily victory or defeat.
>
> —Robert Louis Stevenson

What is this book about? How did we come to write it? For whom has it been written?

Like many other readers we have been struck by the recent proliferation of books and films dealing with modern heroes. Distant names from the past—Robert Scott, George Mallory, and Sir Ernest Shackleton—have all been the subject of recent books and Shackleton's adventures in the Antarctic have been the subject of museum exhibits from London to New York. There is also talk of a Wolfgang Petersen film, charting Shackleton's 800 mile journey in a small, open lifeboat to South Georgia Island. Finally, Petersen notes, he could do a sea film (unlike *The Perfect Storm* and *Das Boot*) in which the characters *survive.* Attempts on Mt. Everest, both successful and unsuccessful, are regularly reported in the popular press and two particular attempts have been the subject of a recent IMAX film. The discovery of Mallory's body on Everest in 1999 has sparked additional interest in him and in others, such as Rob Hall, who have lost their lives near the mountain's summit. The recent ascent to the top of the world by the blind climber, Erik Weihenmayer, merited a full cover story in *TIME* magazine, "Blind to Failure" (6/18/01).

This interest in heroes is particularly compelling since we hear so much in the contemporary popular media concerning the relative absence of heroes.

One of the answers that is often given is that the media themselves have stripped us of the possibility of heroes. When we learn every sordid detail of every public figure's life, the argument goes, we lose faith in the possibility of heroism. Some have even denied the existence of *bona fide* heroes and heroines, believing that if the full facts were revealed we would see that all putative heroes were really not heroes at all, but rather frail, flawed human beings like the rest of us.

How should we respond to these dual phenomena—the apparent prevalence of heroes or at least prevalence of interest in possible heroes and the attitude that these individuals were probably, at base, all too human? Should the latter come as a surprise to us? Should the existence of human failings disqualify individuals as possible heroes?

Our interest in this subject began with a set of observations and attitudes that emerged from our reading, our experiences, and our classrooms. First, while we can understand the current skepticism regarding heroic achievements we do believe in the possibility of human heroism. Examples have existed from the dawn of human memory to the present time. The behavioral and professional categories from which heroes have been drawn may change, but the identification of heroes persists. We have also been struck by the fact that many of these individuals—from antiquity to the present, in fiction, myth, and history—have carried wounds of various kinds, from the physical wounds of the epic hero to the developmental wounds identified and highlighted by modern psychology.

We have wondered if the wounded hero is more the rule than the exception. Is there, in fact, a relationship between the wounds which the hero carries and the accomplishments which he achieves? If so, are there important lessons that might be drawn from the hero's experience?

In studying the approaches and conclusions of modern psychology we have been struck by the extent to which developmentalists have envisioned human life as proceeding through a set of predictable stages. In *As You Like It* one of Shakespeare's characters, Jacques, delivers the famous "7 Ages of Man" speech. Modern developmentalists have also followed this tack, the result being that expectations are raised concerning each individual's negotiation of these stages (the number and nature of the stages varying from theorist to theorist). When a person is "expected" to pass through certain stages in a certain order, norms are established. Even if these are reasonably flexible, patterns become models and individuals whose lives do not follow these patterns are seen as abnormal. To psychologists, as with their friends and neighbors in contemporary society, *abnormality* is seldom considered to be a good thing.

We have been considering these issues over the last two decades and have tested our tentative conclusions on multiple student groups. These groups have consisted of individuals ranging in age from 18 to 85. We have benefited from the experiences, opinions, and insights of our students and have added them to the mix of materials that have helped to form our conclusions.

From this set of observations and conclusions came the argument of this book: if our heroes are often developmentally "abnormal" and if our heroes serve as our *models*, should we not adjust our sense of authentic and meaningful human experience and pay a bit more attention to the *exceptional* and a bit less attention to the "normal"? The "normal" are seldom our heroes. Why then should they serve as our *models*, i.e. the models proposed to us by developmental theory?

The obvious next step was to inquire concerning the possible relationship between the "abnormalities" and the exceptional accomplishments of the heroic personality. Are there lessons which we can learn? Are there ways in which we can even be inspired by the hero's or heroine's experience? Are there pitfalls that we can identify in their lives and, hence, avoid in our own? The answers to those questions add nuance and detail to the book's central argument and offer us further opportunities for thought and reflection.

Thus, the book assembles a number of materials and observations and attempts to serve a number of purposes. It uses history and biography to explore the experience of a number of historical heroes; it uses the texts of popular culture to highlight certain aspects of fictional heroes; and it uses the insights of developmental psychology to elucidate the importance of "stage" theories of human life. In combining these materials we hope to offer information and insight, encouragement and inspiration. The book argues that our views of human development should be adjusted in light of the experience of our heroes and heroines and it suggests ways in which we might contextualize our own experience and move beyond it to personal heroism.

Hence the book is designed for self-help and awareness as well as for personal inspiration. While it argues for some adjustments to current developmental thought, it deals with human "wellness" broadly conceived. It is not, exclusively, a scholarly book, but there is scholarship behind it. It is a book that uses certain scholarly materials to address issues of broad human interest for a broad readership. While we hope that it is engaging and thought-provoking we have attempted to make it as accessible as possible to all. It might be thought of as a self-help book designed for readers who are interested in psychology, in history, in literature and in film, readers who are aware of their own 'wounds' but who wish to move beyond them and who might take heart in the experiences of others who have done so.

What is a human life? Of what does it consist? How should we think of it?

In the Venerable Bede's (ca. 672–735) *Ecclesiastical History of the English People* there is a memorable passage which some consider the finest in Old English prose. The passage concerns the experience of religious conversion and it is built upon an image designed to capture both the meaning and mystery of human life. The following is a loose translation:

> Another of the king's counselors and chiefs took up the discussion: 'It seems to me, my liege, that man's life on earth might be described in this way--a king sits at dinner with his chiefs and retainers in his meadhall. It is the depth of winter and a fire has been kindled to warm the hall. Outside it rains and snows and storms. Suddenly a sparrow appears within the house. It has flown in through one door; it flies through the warm hall, and flies through another, back into the darkness. So it is with man. His time here is brief. What went before and what follows after we know not. If there are words and beliefs which might help us to understand all of this, we should attend to them.'

One explanation for the experience behind this analogy comes from faith, but faith is a gift and all do not share it. Faith is not, as the eighteenth-century philosopher David Hume argued, subject to rational argument since it is, by its nature, beyond reason. Even those who believe in or posit a future and/or a prior existence are unable to describe that extra-temporal past and future with any specificity.

The king's counselor imagines the existence of such states but wraps them in a darkness impenetrable to human experience and reason. The life which we know, the life which we can study, is bounded by two great events: the certain fact of our birth and the certain fact of our death. As rational beings we contemplate those events endlessly. Birth and death help to define us because of our consciousness of them. We confront those realities at a level of concern and understanding that the birds and beasts and insects do not share. Birth is easier. It initiates us into life. It brings us into a world of possibility. We believe we understand its cause and, to some degree, its purpose. That life into which it initiates us is more problematic; we do not understand its purpose at the level of comprehension which we seek. Death is more difficult still. It is our horizon; it delimits us and, hence, shapes us. No matter how hard we try to follow the advice of the poet William Blake and attempt to see through and beyond that horizon, it looms before us as a central certainty in our life.

Presented with these essential, eternal certainties we see our life as a story, a story with a beginning, a middle, and an end. The fact that we cling to that life so dearly is evident in the measures we use to mark its stages. The "middle" years come far closer to the end than we wish to acknowledge and we demonstrate the value we place on the intensity of the experience by constructing and supporting a system of formal education which occupies nearly one-third of the projected period in which we expect to occupy the earth. Life is brief but our preparation for much of it is long.

Aristotle is not the greatest literary critic in the west solely because of his ability to describe and analyze Greek drama. His insights on the drama are insights on human life, for the drama, as he argues, is an imitation of an action and that action, ultimately, is life itself. All of our popular entertainments, our films, television dramas, and genre fiction, are built upon Aristotle's notions and every guide to writing such works are recapitulations of Aristotle's *Poetics*.

The *ur*- or original story, the fundamental plot, concerns a decent but imperfect individual whose life is rent by conflict but, eventually, culminates in a climax and is ultimately resolved in such a way as to provide us reassurance. It may reassure us by ending happily or it may reassure us by embodying hard-won wisdom. In either case, as the great critic, Samuel Johnson, argued, it is a story whose arc and details carry both novelty and truth. The story is predictable in important ways. How could it not be, since each individual life recapitulates the arc of all such lives. At the same time, it is fresh and new, for each individual life is unique and, thus, surprises us, even as it teaches us lessons we have long known.

We see our lives as such stories and their predictability is essential. Each of us is the protagonist of that action, the story's hero. Each of us (we believe) is decent but flawed. Each of us experiences conflict and each of us (realizing the necessity for an ending) longs for a proper one. In this age of social science and statistics we desire the predictability which those statistics suggest we will enjoy. Death does not disturb us; early death does. We speak of a "full life," not an endless, earthly one. All children will die, but they should not die before their parents do. To everything there is a season and we wish our lives to be similarly charted, though we are concerned that they also be neither tedious nor dull.

Developmental psychologists share this wish and they have constructed systems for marking the scenes in our lives, the "stages" of our experience. The fact that they have done so provides further reassurance for us, the actors and actresses, who, as Shakespeare promised, will play many parts in our time. We are comforted by the knowledge that our feelings and responses in the face of difficult events—the plot points of our lives—are neither singular nor aberrant. We are reassured by the fact that that which we have suffered has been faced and survived by others and that our feelings, traumatic in their apparent novelty and intensity, are, in fact, eternal and quintessentially human.

At the same time we realize, with Job, that there are false comforts and false comforters. We might seek to extend our lives by attending to our health and diet and behavior but we continue to take risks in our search for a greater good. The classic Stoicism of antiquity, for example, may counsel us to avoid taking a spouse and having a child for they will be as fragile as clay vessels and put our peace and equilibrium at risk, but many of us choose to ignore this ancient advice. In spite of the later warnings of the scientist and essayist, Francis Bacon, we regularly give hostages to Fortune. We build homes on fault lines and in the shadows of volcanoes; we compete in contact sports; we explore mines and caves and we travel in both earthly and extra-terrestrial space.

It comes as no surprise that we should do this, for life, like Greek drama, is built upon conflict. Without danger, without risk and challenge, there is no interesting action. Predictability, in short, is not always seen as a virtue. At the same time, however, we tend to avoid conflict involving high and potentially, mortal risk. That is why the "sublime" painting of the eighteenth century was so popular. It depicted scenes of great danger and threat (devastating storms at sea, for example), but from a vantage point where the viewer felt safe. Observing a distant storm may pique our imagination, but if the distance is reduced we feel fear rather than awe.

While there are some readers who prefer writing that depicts unrelieved horror and some who prefer writing that offers endless pages of maxims and apothegms, the great majority of us prefer writing that both stimulates and comforts, that offers surprise as well as predictability. Our lives are sometimes dull and we seek excitement and stimulation but our lives are also sometimes stressful and we seek comfort and reassurance. Thus, there is no contradiction in the fact that we sometimes feel most fully alive when our abilities are significantly stretched while at other times we feel most fully alive when we are in a state of complete peace, freed of all concerns.

It has been said that vacations and retirement are the natural state of man. It is difficult, however, to appreciate personal peace without having some experience of its alternative. The philosopher William James argued that we should attempt to multiply the holidays of life, but he realized that the holidays were defined in good measure by the days that separated them. Leisure without challenge is like challenge without respite. Neither offers that balance which humans seek both in their art and in their lives.

We mark our lives through the use of elaborate verbal and iconographic systems. We have ranks and titles, medals and badges, daily and special-occasion uniforms. We receive promotions that are accompanied by real emoluments and symbolic trappings, each of which is seen as important. Human resource analysts divide the jobs we do into categories with multiple divisions and steps. Our walls are decorated with diplomas and plaques, certificates, and record-shot photographs. Church liturgies and sacraments give both cyclic and linear shape to our lives. Our lives are punctuated by key ceremonies with accompanying music, dress, speeches, and greeting cards.

At the same time there is something oppressive and cloying about all of this. We sometimes characterize it as a "game" and often see it as inauthentic. We seek escape from the world of the time-server and the ticket-puncher. We value energy and individuality over order and the same dull round. It is not an accident that the army allows generals to bend the regulations concerning uniforms. Monty, with his outback hat and umbrella, is the perfect counterpart to Patton, with his ivory-handled revolvers. Armored troops wear "tankers' boots," which are equipped with straps, unlike the conventionally laced boots of all other soldiers. While they are in tankers' "uniform," they are simultaneously out of (normal Army) uniform. This is perceived to be one of the distinct advantages of the Armor branch—the opportunity to wear the elements of a uniform that diverges from the *uniform*.

In the pages that follow we will be looking at a critical aspect of this binary dimension of human life. As noted above, developmental psychology has studied the arc of human life, the plot, if you will, of the human story, and it has concluded that life can be depicted as consisting of a set of stages. The stages highlighted might be cognitive (Piaget), sexual (Freud), moral (Kohlberg), psychosocial (Erikson), or some combination of those (or other) dimensions of human experience, but stages are identifiable and they are subject to study.

These stage theories or stage depictions tend to be weighted toward the predictable rather than the unique, toward the "normal," rather than the exceptional. They do not do so naively. It is simply the fact that structural theories or descriptions tend to stress *structure*.

In our judgment, "normality," to the extent that it can be defined and delimited, plays an extremely important role in developmental discourse. At the same time, it must be balanced with the exceptional. Just as we choose "normality" over deviance and pathology, we simultaneously choose boldness over risk-aversion, the sound of the different drummer and the road bravely taken over the conventional melody and the common rut.

This is a book about heroism and it is the case that many of our heroes have led exceptional rather than normal lives. It is also the case that many of these heroes have struggled with what might be termed developmental disabilities, i.e. disabilities *vis à vis* the expected pattern of "normal" experience. We find inspiration in their experience and believe that, like them, we ourselves might be able to both transcend our suffering and—as they often have done—build upon it. We might, in short, seek our personal heroism in our ability to capitalize upon our "abnormality," which is to say, our individuality.

To return to the notion of human life as story, we know that we are flawed but hope that we are decent, caring, and, as was once commonly said, *authentic*. At the same time, we understand that the flaws in our lives contribute to our identity. They are a part of us. In some cases we seek to outgrow them or transcend them. In others—the subject of much of this book—we seek to build upon them. Our experience is *our* experience; it constitutes us. Heroism is, in a sense, the full utilization of that experience, the full development of the individual personality and self. We build upon our flaws as well as our strengths and virtues and, with some luck and diligence, those flaws or gaps can enable us to not only succeed but triumph.

The purpose of the discussions that follow is not so much to criticize stage theories as it is to remind ourselves of the extent to which stage theories can be complemented by accounts of the lives of those heroic individuals whose lives have followed other-than-"normal" patterns. We are well aware of the fact that "abnormality" can sometimes lead to pain, suffering, violence, and antisocial acts. It is the case, however, that some "abnormality" can lead to heroism and even serve as heroism's foundation. Thus, our purpose is not to challenge the insights of stage theories but to offer reminders of the other side of human life, reminders that, we hope, can prove useful and inspirational.

The core idea, the principal psychological content and historical examples of the book are Judith's, most of the literary references Richard's. In discussing the materials of the book over several years we have developed a number of the ideas together and in conducting workshops and in team-teaching courses, we have benefited materially from the observations and insights of our many students. To all of them we here express our deep gratitude.

Acknowledgments

Countless family members, friends and colleagues helped, supported and heartened us in myriad ways in the development of this work. To the following mentors, colleagues and friends who have sustained us from the beginning of this endeavor we offer our love and thanks: Dr. Michael J. Collins and JoEllen Collins, Dr. Phyllis O'Callaghan, the late Reverend Royden Davis, S.J., Reverend Robert Lawton, S.J., Dr. Jesse Mann, Dr. Norma Small, Dr. Sherry Eve Penn, Dr. Harry B. Green, Dr. Luisa Contreras, the late LTC Walter Pritchard, Maryellen Pritchard Beech and CPT Wayne Beech, the late Dr. Donald Morgan and Susan Morgan, Dr. Carolyn Loram and the late Dr. Ian Loram, Patricia and Larry Veleke, Dorothea Eichhoff, Jamea French and Dr. Larry French, Katy and Charlie McCaffrey, Jude Ludlow, Noreen and Gary Greve, Barbara Karavolas, Alice E. Benn, Nancy and John Spaeni, Dr. Gordon and Karen Tuffli, the late Sandra Lucy Campo and Alfredo Campo, Claire Fallon, Robert Erickson, Elaine Bradbury-Stewart, Carrie James, Mary Harris, Dawn Stanton, Mary Ann Capano, Heather Hanford and Jim and Marian Foster. To them we will always be indebted.

We offer thanks to a number of very special friends and colleagues, in particular, the late Reverend Timothy S. Healy, S.J., Pamela Bianculli, Theresa Coughlin, Rogelio Fussa, CPT Michael and Sherry Murphy, Rebecca Staley and the late Larry Staley, Jennifer Hunt, Julie Ito, Elvira Roberts, Anne Ridder, Dr. Mary E. Bussell, Mrs. Susan Bussell, Dr. Ted Tarkow, Jill and David Housh, Barbara Staiger, Linda Moeller, Traci Wren, Nancy Moen, Anne Weller, Mike Watson, Bill Sheehan, Melanie and David Johnson-Moxley, Dola Haessig, Danny Glover, Melody Galen, Ellen McLain, Lindsay Young Lopez, Dr. Donald and Mrs. Lona Packwood, the late Col. Arthur C. Allen, Gary Tatlow and the late Marilyn Tatlow, John Marshall and the late Betty Marshall, Richard Miller, Stuart Fraser, Andrew Sackin, Jim and Micki

Amos, Mrs. Dora and the late Jim Fergason, John and Cynthia Shaw, R.D. and Loretta Ross, Kent and Dottie Kreh, Walter and Beverly Pfeffer, Dennis Barnes, the late Mary Nell Porter, Dr. Kenneth Mares, Beverly Skyles, Professors Thomas DiLorenzo, Harris Cooper, David Geary, Ann Bettencourt, Dr. Gary Smith and Carol Smith, Drs. Richard and Elaine Hocks, Dr. William Bondeson, Tim Rooney, Dr. Lori Franz, Chancellor-Emeritus Richard Wallace and Mrs. Patricia Wallace, Chancellor Brady Deaton and Dr. Anne Deaton and Dr. Brian Foster and Mrs. Lerke Foster.

To our colleagues, friends, learners and facilitators who kindle the spark of learning throughout the life-span, we give our profound thanks: at the University of Missouri-Columbia, Dr. Lucille Salerno, Laura Ward, and Adam Newman and at Washington University in St. Louis, Rich and Margaret Diemer, Betsy Gee and the late David Gee, Marion E. Horstman, Henrietta and Rudy Freedman, Frances Franklin, Anne Hetlage, Jane Smith, Charles Moore, Jackie Ess, Drs. Bryce and Betty Hudgins, Robert and Dorothy Shores, Nancy McKee, the late Richard J. Connors, Anabel and Wayne Hudgins, the late Larry Jasper, Les Gaffner, Fred Ashner, Dr. Leon McGahee, John Lesser, Richard Hyde, David Crowley, Dorcas Case, Dr. Richard Sisson and Mitzi Sisson, Stuart and Becky Haw, Brooke and Florence Hoey, Bob Hollander, Jesse Selvey, Corinne Richardson, Dolores Friesen, Betty Schmelzer, Philip Rogers, Dr. J.J. Burke, Dr. Jane Miller, Sue and Marty Wohl, Jane Elswick, Richard Klein, Dr. Lloyd Klinedinst, Barbara Bollmann, the late Sam Goldblatt and Gloria Goldblatt, Patricia Barrett, Sandy Christie, Tom Treeger, Sue Clancy, Sol and Barbara Linder, Jackie Kasnetz, Dr. Ben Milder and Mrs. Jeanne Milder, Jerry Kottler, Norman Solomon, Joan Dellbringge, Drs. Bob and Mildred Kaufman, Jane Weinhaus, Allen and Lee Furfine, Ed Koch, Dr. Ben Borowsky, Ron Riess, Dionne Grant, Drs. Alan and Elizabeth Korte, Charles Dignan, Robert and Maggie Bauman, Ruby Quarterman, Don Godiner, Joan Burtelow, Mike Pinney, Bob Schwartz, Jim Steffen, Nancy and Ed Schapiro, Vivien Becker, Tom Hahs, Gladys Barker, Gale Murphy, Ben Sandler, Alyn Essman, Barbara Lawton, David and Dee Felt, Laura Blumenfeld, Herschel Asner, Dr. Milica and Stanko Banjanin, Dr. Larry and Jane Kahn, Karen and Butch Sterbenz, Val Levi, Sally Kopman, Louise Lonsbury, the late Marjorie Berg, the late Connie King, Barbara Ritchey, Catherine Rankovic, Maria Hunter, Carole Wallace, Diane Willis, Rick Schneider, Carol Murphy, Pat Agnew, Arika Cannon, Erin Presson, Britt Congemo, Katina Truman, Dr. Rebecca Copeland, Dr. Steve Ehrlich, Brent Jenson, Katie Compton, Dean Robert E. Wiltenburg and Chancellor Mark S. Wrighton.

We have received exceptional assistance from both the past and present staff of Westminster College's National Churchill Museum in Fulton, Missouri and it has been a true pleasure to work with all of them: Dr. Gordon

Davis, Dr. Jerry Morelock, Warren Hollrah, Lori Conner, John Hensley, the current Executive Director, Dr. Rob Havers, Sara Winingear, Assistant Director, and Mandy Plybon, Education Coordinator. The wonderful and dedicated members of the Advisory Board of the Mid-Missouri Friends of the National Churchill Museum who tirelessly give of themselves (Genevieve Connor, Cathlynn Dodson, Jane Forsythe, Mary Harrison, Susan Krumm, Debbie LaRue, Nancy Lewis, Whit McCoskrie, Jody Paschal, Barbara Stassel, Eva Sterner and Joyce Williams) have always provided outstanding support, in particular the late Helen W. Danuser, who was witness to Churchill's speech in Fulton and who served as the first President of the Advisory Board and the first Museum docent. We also wish to acknowledge the kindness of Suzanne and David Richardson. All of these individuals have helped to preserve and extend Churchill's legacy.

We have been privileged to receive the advice and encouragement of Churchill's daughter and granddaughter, Lady Mary Soames and Celia Sandys, as well as that of Colonel Eric A. Orsini, who fought with Patton's Third Army and served for many years as Deputy Assistant Secretary for Logistics, Department of the Army. To them we give our sincere thanks.

We would also like to thank our editors, Samantha Kirk and Beverly Shellem, for their advice, support and steady hands. We very much appreciate their efforts on our behalf.

Our deepest gratitude goes to the members of our family, especially our sister and brother-in-law, Jane and Arnold Gibson and our wonderful nieces and nephews. Our son, Jonathan, daughter-in-law, Kirsten, and Katharine and Caroline, our granddaughters, who each and every day enrich our lives and enliven our spirits are a true and constant blessing and inspiration. Our parents did not live to see this work come to fruition but their presence and support is always profoundly felt.

Yeats's famous lines ("Think where man's glory most begins and ends,/ And say my glory was I had such friends") apply to us. Those friends' names are recorded above.

Highland Woods
Columbia, Missouri
November, 2009

Chapter I

Introduction

If everybody was satisfied with himself, there would be no heroes.

—Mark Twain

Where do we find these wounded heroes? Who speaks for the "normal" or common and why? Do we find our heroes or are they created for us? How has scholarship shaped this discussion?

The great European comparatist, Erich Auerbach, begins his celebrated study of the representations of reality in western literature, *Mimesis*, with an account of Odysseus' scar. In book 19 of Homer's *Odyssey*, after years of travel following the Trojan War, Odysseus has finally returned to his home in Ithaca, where his queen has been subjected to the appeals of would-be suitors. Rather than immediately announce his presence he attempts to pass in disguise and presents himself as a stranger seeking hospitality. His wife Penelope (who has not recognized him) instructs Euryclea, an old house-keeper, to wash the stranger's feet. At that point Odysseus realizes that he will be instantly recognized because of a scar which he has long carried. He is correct; the moment that Euryclea touches the scar she permits Odysseus' foot to fall into the basin as she is overcome with joy.

Auerbach is struck by the manner in which Homer then retells the events which produced Odysseus' scar. The scar was actually received in boyhood, when Odysseus was participating in a boar hunt with his grandfather Autoly-cus. The realistic depiction of the hunt and the resulting scar from the boar's tusk add subtle layers to Homer's narrative.

For most readers, however, the most immediately noticeable aspect of the passage will be the fact that it is a scar which identifies the hero and not some indication of genius or exhibition of prowess. For Dante, Odysseus (or Ulysses) was not a hero but rather a person to be consigned to hell for his counseling of fraud. In addition to his reputation as a traveler and adventurer, Ulysses was known for being wily and crafty. He was, after all, the principal perpetrator of the Trojan horse hoax which led to the fall of Troy. It is this aspect of Ulysses which seizes Dante's attention. For Dante, those who counsel fraud belong in the infernal regions. For Tennyson, on the other hand, Ulysses is the indomitable traveler, the individual who would not be conquered by age but who would continue to seek new worlds even as he reflected on the worlds of whose histories his own experiences were now an inextricable part. Anyone who has ever sat through his or her share of after-dinner speeches has heard Tennyson's words intoned, usually in connection with some challenge to the assembled membership: "to strive, to seek, to find, and not to yield."

For Homer, Odysseus is many things, but that aspect of his experience by which he is identified—identified, we would emphasize, by those who know and love him best—is his scar. Wounded heroes are common in mythology; we know many individuals through their weaknesses and use those weaknesses to identify them. Achilles, the principal hero of the *Iliad*, is a common example. According to legend, Achilles' mother Thetis attempted to make him immortal by dipping him in the river Styx. She succeeded, with a single crucial exception: she held him by his heel and was unable to immerse the heel in the river. The heel became his single point of vulnerability. Now, the phrase "Achilles heel" has passed into common parlance as an image of a fatal weakness in an otherwise powerful figure or force.

Throughout both mythology and the history of religion kings and gods die for their people; in their weakness and mortality lies a crucial portion of their strength and in their passing we see an image of our own development as individuals and as members of societies. The old are replaced by the young, the weak by the strong; as Ecclesiastes would have it, there is a proper season for all things, a time to sow and a time to reap, a time to be born and a time to die.

The heroes we will encounter in the discussions that follow were also wounded in special ways. We will be particularly interested in those wounds which can be seen as parts of (sometimes fragmentary parts of) developmental processes. We will use the experience of the wounded heroes and heroines to help us to refine and enrich some of the conclusions of developmental theory. We will argue, in brief, that while developmental theory sees the healthy, happy human being as a person who has successfully passed through a gauntlet of experiences in a predictable, carefully-prescribed order, the hero often fails the exercise but survives to achieve much more. In his experience is our lesson. The corollary questions follow: if our heroes and heroines can live with their scars, how can we? What are the lineaments of heroism and how distant are they from our own capacities? Should our personal wounds serve as the bases for new opportunities or as the justification for self-pity?

Discussions of human development tend to focus on patterns of behavior which are perceived to be both *orderly* and *complete*. In this they mirror the medieval 'great chain of being', a descriptive metaphor for the cosmic order, with its twin characteristics of hierarchy and plenitude. The 'great chain' notion was finally discarded after the successful attacks on it of Enlightenment figures, who reasoned that it was not as 'complete' as it pretended to be and that its concept of order and hierarchy gave rise to visions of society and the social order which were not carved in granite but rather subject to human will, effort and desire.

For the strict stage-theory developmentalist, fulfillment is achieved by passing through a specified series of stages in the right ways and at the right times. Heroes, on the other hand, often shatter these patterns, carrying, ever after, their developmental disabilities or wounds. All of us carry such wounds, though we might not think of ourselves as heroic. Sometimes they scar over and are forgotten; sometimes they persist and are sources of great pain. For the hero they can be occasions of triumph. In some cases their role in the hero's triumph might even be said to be causal, the wound leading to the achievement. How does one use one's wounds as a source of healing rather than one of lingering pain? How does one turn one's wounds to personal and public advantage?

The hero may fail to move from one developmental stage to the next because of some trauma or great personal challenge. He or she may, indeed, have halted at a single stage indefinitely or even skipped one or more of the expected stages. We would argue that our analysis of the heroic personality should include the insight that heroes are often those who succeed in spite of their inability to negotiate certain developmental stages. Indeed, these heroes may have succeeded *because of* their psychic wounds, wounds upon which their heroic character was eventually built.

Heroes were once considered to be descended from gods, but the historical heroes which we will discuss were quite human. They suffered from experiential gaps, personal disabilities, and even from deep human flaws. They are heroic models for us for many reasons, but one is their ability to triumph in the face of pain, loss, and apparent weakness. They do so, we will argue, not because they can simply *survive* experiential and developmental gaps. They do so because they possess the *virtues* associated with specific developmental processes (on which more later), virtues which might be considered to be essentially synonymous with the heroic. The resulting corollary is that we should focus equal or greater attention on the developmental *virtues* than we do on the developmental *stages*. The former are a *sine qua non* for heroism; the latter are not. Put another way, character is finally far more relevant than the stages or "passages" in one's life. *Character is essential; a tidy pattern is not.* The point is not that pain builds character (a favorite belief of the Victorians), though it can. The point is that character can be built and pain transcended.

Those of us who acknowledge that we admire individuals who we consider heroes or heroines (some of our students have told us that they do not have *any* heroes) might, in our personal and familial lives, subscribe to a psychology of *normality*, a psychology in which the negotiation of stages is of central importance for human happiness. We can be forgiven for that, since it is difficult not to see life in those terms, for they are implicit in nearly every cultural, religious, and economic structure which surrounds us in our daily lives. It is striking, however, that in our admiration of the heroic we never seem to single out the *hero's* "normality" as a characteristic of his or her heroism. His or her abnormality—the manner in which he or she has transcended circumstance or made a virtue of a perceived disability—is what seizes our attention. Hence, while each of us might strive, in some fashion, to be "normal," the hero holds out hope to us that any potential failure in this endeavor need not prove fatal. On the contrary, the hero's life suggests to us that out of weakness or absence or loss might come not only achievement but even greatness. This is, in part, a source of the hero's capacity to *inspire* and that capacity is far more important than any orderly pattern of experience.

The notion of "normality" as a benchmark or guide has been much discussed. Offer and Sabshin (1984), who have argued for the development of a research area which they term "normatology," use the word "normality" interchangeably with "mental health." Psychoanalysts, however, have often followed Freud in arguing that neurosis is more common than "normality" and that "normal functioning" is an ideal and, hence—in daily life, as led—a fiction.

Normal is from the Latin *normalis*, which suggests that something has been "made according to rule," that it conforms "to the standard or the common type" (Offer and Sabshin, p. 364*n*). Thus, behaviorists tend to see "normal" behavior as "average" behavior. The possible negative implication is that those who are normal may also be described as "middling" or mediocre, while most of us want to be "above average," like the residents of Garrison Keillor's Lake Wobegone.

Developmentalists have taken a different tack from the psychoanalysts and behaviorists, imagining, by and large, a world in which humans can aspire to health, happiness, and self-actualization. Neurosis is not the norm and "normality" is purged of unfavorable associations, either through its specification as an explicit measure or through its submersion within the model as an implicit one.

Nevertheless, it has long been recognized that abnormality can be a gift as well as a burden. In his dialogue, the *Phaedrus*, for example, Plato points out that there is a madness that is evil and there is a madness which is a divine gift. The prophets, for example, are abnormal—some would say, mad—and they are most fully themselves in that madness.

Coleridge imagines such a state of inspired delight in his poem, "Kubla Khan":

> A damsel with a dulcimer
> In a vision once I saw:
> It was an Abyssinian maid,
> And on her dulcimer she played,
> Singing of Mount Abora.
> Could I revive within me
> Her symphony and song,
> To such a deep delight 'twould win me,
> That with music loud and long,
> I would build that dome in air,
> That sunny dome! those caves of ice!
> And all who heard should see them there,
> And all should cry, Beware! Beware!
> His flashing eyes, his floating hair!
> Weave a circle round him thrice,
> And close your eyes with holy dread,
> For he on honey-dew hath fed,
> And drunk the milk of Paradise.
>
> (ll. 37–54)

This poetic vision could be linked with the notion that creativity is somehow dependent upon neurosis (or drug-induced visions), an idea which arises frequently, particularly in discussions of the reputed fine line between genius and insanity. While it is true (as Freud argued) that creativity is sometimes the result of deflected energy (Yeats once said that if he had been happy he would have lived rather than written), it is hardly the result of insanity, for, as Samuel Johnson argued, true genius is closely allied with *sanity*, not its opposite. Plato comments that the young men of Athens sometimes stopped bathing because they thought it would help to make them like the poets, but the writing of poetry requires far more than the absence of soap. The fact that one attempts to drink like Hemingway or Faulkner does not mean that one will also be capable of writing like them.

While not embracing a purely behaviorist model, we would argue that the favoring or privileging of "normality" is open to challenge and qualification. Similarly, while not embracing a purely psychoanalytic model, we would argue that the developmentalists' (often unspoken) belief in the possibility of "normality" as a common course rather than an ideal also requires further study and reflection.

The "normal" is not necessarily dull or mediocre, nor is its experience hopelessly rare or impossible. The problem is that its attributes can be excessively praised while the attributes of its opposite can be insufficiently appreciated. All are in agreement that any depiction of the "normal" is both culturally and temporally conditioned. Anthropologists who examine the variations among practices and beliefs across the globe regularly demonstrate to us that *our* "normal" is not necessarily *the* "normal." Given that fact, we should be careful not to privilege it excessively. Given the parallel fact that *each of us* to some extent experiences the "abnormal," our sense of the "normal" should be adjusted, particularly in light of the degree to which we see abnormality in our heroes and heroines, individuals who are, in important ways, our *models*. The "normal" state of the human body should include freedom from disease, but each of us experiences disease from time to time and we respect individuals who triumph over the diseases with which they have been afflicted.

The "models" which we often select are seldom, in the common sense of the term, "normal." This is not to say that our heroes suffer from some pathology. The hero's form of "abnormality" stands in parallel to developmental theory, not contradicting it so much as adding nuance and subtlety to it, footnoting and qualifying it, balancing it. Developmental theory must be adjusted to accommodate heroic experience just as heroic experience can sometimes be contextualized as a succession of achievements that are, in certain instances, precipitated by disabilities. Moreover, this inclusion of the heroic perspective extends the notion of "development," suggesting that rather than seeing a "normal" progression through stages as a complete progression we should instead include in our sense of that progression the need to continually expand and extend it—to enlarge one's vision and ask, "why cannot our sense of pattern be richer, deeper, less orderly, and more meaningful?" Why should a sense of the heroic be separated from our everyday lives? Should not the more indirect, gap-laden but yet inspirational pattern of the hero's life be available to all of us as a possible norm? Should we not all be permitted to aspire to heroism? Should not our own wounds be seen as part of the fullness of our *individuality* rather than as a list of behaviors (or lost experiences or chances) through which we diverge from the norm?

We will look at a number of heroic personalities and devote particular attention to four principal examples: T. E. Lawrence (Lawrence of Arabia),

Winston Churchill, George S. Patton, Jr., and Jane Austen. These four individuals have been chosen for several reasons. In the first place, they have achieved recognizable prominence, a prominence that justifies, in some measure, the construction of generalizations based on their experiences. They are each major 'media' figures, whose stories have been told and retold in multiple formats (biographies, novels, 'children's books', films, and televised narratives). The 'media' prominence of certain individuals indicates the cultural resonance of their life stories. While all would not see them as heroes (Patton in particular draws divergent reactions), a significant number of individuals have seen them either as heroes or, certainly, as interesting and compelling people. Moreover, their experience closely approximates dictionary definitions of the heroic. They have also been seen as models of their kind. Three served in the military. Each was an effective writer. Each came from a personal background of some psychological interest. Each served in a capacity of intellectual leadership which affected the lives of those both within and beyond their personal sphere of influence. Each grappled with individual psychological challenges and each continues to be celebrated today, decades and even centuries after their deaths.

Both Patton and Lawrence have been the subjects of award-winning films. Churchill is an iconic figure, with a military and journalistic as well as political background. He also received the Nobel Prize for literature. Jane Austen's life was far more private, but her fame and cultural impact have never been greater than at this moment. (Bookstores now dedicate entire sections to her and her work, while her novels are repeatedly adapted for cinematic or televised presentation.)

Their role here is purely illustrative. We might have selected hundreds of other individuals for consideration. We chose people with whom we could assume a certain degree of public familiarity, people in whom we had an interest, people whose lives and personal development, we thought, would be of interest to others.

They do not represent "cases" *per se* in the sense that they embody syndromes that should be seen as in some ways archetypal. The general discussion of heroism which follows in the next chapter is designed to lead the reader to think of his or her *own* examples of the heroic. The discussion of the life stories of particular individuals is designed to stimulate the reader to pursue the biographical details of the lives of his or her own heroes and to assess the degree to which our analyses and descriptions are useful in studying the lives of other individuals beyond those we have space to discuss here.

If we are to see life whole we must be prepared to draw materials from varying quarters of our experience. In the case of the individuals mentioned above one must deal with materials from such areas of thought and experience as developmental psychology, history, biography, film, popular culture, and heroic theory. Each of those areas (and the list is hardly exhaustive) serve to construct and contextualize our own experience. Thus, in the pages that follow, we will move freely, back and forth among these areas.

The importance of heroes in human history is easy to demonstrate, both in the results of their actions and in the concurrent and later representations of those actions. In our own time, earlier but continuing forms of representation (oral and printed accounts, portraiture) have been combined with new media (cinematic, photographic, electronic) to construct images of heroism for broad popular consumption. In the twentieth century a literature—both popular and scholarly—has developed which both categorizes forms of heroism and representations of heroism and also attempts to assess the human need for heroes.

With modern communications, however, there is an increasingly important interplay between desire, need, and representation. We need and desire heroes; hence the media provide us with heroes. The media are successful to the extent that they are sensitive and responsive to our needs and desires. At the same time, the nature of the representations which they offer us helps to shape those needs and desires. David Lean (*Lawrence of Arabia*, 1962) chose to make a movie of the early life of T. E. Lawrence for a number of reasons. One of those reasons would be his expectation that members of the general public would wish to see such a film and would pay money to do so. Having given us such a film, however, he has enriched our perceptions concerning Lawrence. Since its production, however, the common, popular conception of Lawrence has tended to focus on certain aspects of his life (those stressed by the film) to the exclusion of others. Lean's Lawrence, to a considerable extent, is now "our" Lawrence, a Lawrence without youth, adolescence or middle age.

Thus, we get what we want (and deserve?) from the media, but the media also attempt to shape what we want. Given the propensity of both publishers and filmmakers to attempt to repeat previous successes, we also tend to get more than we want of certain types of representations. When a film is particularly successful the next season brings a host of repetitions and imitations, all of which hope to capitalize on the market tapped by the original successful production. Hence, a market success creates cultural currents and waves whose long-term importance can be considerable.

This is why we have argued that the particular subject which we will be investigating necessitates the use of varied material, material which is generally not utilized by students of the heroic, whose foci tend to be discipline-based rather than culturally-based. While the literature of heroism has grown and developed in the twentieth century, the literature of developmental psychology has grown simultaneously but separately, despite the fact that their conclusions can be effectively combined. Students of human development are far more likely to be versed in other areas of psychology than in parallel developments in the humanities. Similarly, students of myth, history, literature, and popular culture are more likely to be versed in parallel or complementary fields in the humanities than they are in psychology. When they do introduce psychological concepts in their work they are far more likely to use specific, traditional, theoretical materials (Freud, for example, and, more recently, Lacan) than the conclusions of developmentalists. There is a view of the "normal" in scholarship as well as in life and a sense that pursuing it will somehow result in intellectual health and success. Thus, scholars pursue the practices which they observe in the works of other scholars, doing what Thomas Kuhn calls "normal science."

From time to time developmental psychologists have studied the experiences of famous individuals who, for many, are heroic figures. Erik Erikson, for example, wrote books on Martin Luther (1958) and Gandhi (1969), but most often they have studied common cohorts of everyday people and the methods they have employed have been the methods of the more quantitative social sciences. Thus a contemporary developmental psychologist might ask the question, "At what age do left-handed siblings in Bengal commonly begin to distinguish between consonants and vowels?" rather than a question such as, "What is the cause of civilization's discontents?"

As a discipline, psychology stands at the intersection of the sciences and the humanities and the early leaders in psychology (Freud, Jung, Adler, et al.) were far closer to the humanities than are contemporary psychologists. It is often noted, for example, that Freud learned a great deal from Shakespeare and other literary figures, whose depictions of human life and behavior are replete with psychological insights. In this book we are attempting, however modestly, to invoke that earlier tradition.

Contemporary humanists, on the other hand, do turn more often to public and historical personalities than do psychologists, but in doing so they have been more concerned with representations of those individuals that have been influential because of their aesthetic success and/or the broad public audience those representations have attracted (or, increasingly, the cultural damage they might have done). In studying such figures they are particularly interested in the ideological and political implications of the artists' representations. Shakespeare is seen as a part of the cultural hegemony exerted by Elizabethan elites and historical figures are studied in the interest of giving heart to the disenfranchised and downtrodden or preparing the intellectual basis for a reconsideration of the political structures that resulted in such victimization. Still, it is important that we attempt, in some small way, to bridge both psychology and humanistic cultural studies, since both are so close to our personal experience and relevant for our views of our selves and our societies. Our desire is to draw conclusions from the experience of particular *individuals* in such a way as to give heart to *all* members of the human family and provide part of the basis for a process in which individuals might all shed their victimhood and progress toward heroism.

While we live our lives holistically we study aspects of our experience through the lenses of separate disciplines. Painters draw the human body; physicians dissect it; psychologists analyze, historians report and interpret, sociologists describe and quantify and economists and political scientists theorize and predict. As we look at heroes and heroism we will need to conjure with a multiplicity of facts, needs, and phenomena. In the first place there is a vast array of possible heroic models, not just heroes from the past and the present, but heroes from poetry, film, and fiction. Comic-book superheroes, for example, are often known to a wider audience than the living or dead heroes of human history (though it is interesting to note in the context of this book that many of these superheroes are developmental wrecks. Many found their superhero identity as the result of some catastrophe and are now known by some trait, capacity, or physical feature which both defines them and sets them apart from "normal" human beings).

Heroes fill certain psychological needs and we will attempt to trace those needs. In the process we will define the heroic and attempt to elucidate its more common characteristics. We live in a world in which heroes and pseudo-heroes are constructed by the media, a world in which specific facts can either be highlighted, diminished or wildly exaggerated. We will need to distinguish between authentic heroes and celebrities, the heroes and heroines of the moment who often embody values that are anything but heroic in the traditional sense. As such individuals seek their fifteen minutes of fame predicted by Andy Warhol we must be careful to distinguish their *notoriety* from the accomplishments of true heroes and heroines.

Our ultimate purpose will be threefold. We will try to add nuance to the generalized vision of developmental psychology and the manner in which it foregrounds and privileges notions of normality. We will mount this friendly challenge through the use of the heroic experience and personality and its capacity to extend and deepen our sense of daily human life. Finally, we will suggest that heroism is not some remote set of achievements and behaviors beyond normal human capacity, but rather an opportunity for inspiration and understanding, a way of thinking about human life that invites us forward, suggesting new opportunities both for our lives and for our sense of ourselves. The pains and disjunctions that characterize our lives need not deter us from seeking something beyond normality. None of us need be suffering victims when we can aspire to heroism.

We ask individual readers of these pages to consider and reflect upon the experience of other (heroic) *individuals*. While generalizations can be made about groupings of human beings we believe that some of the most powerful lessons to be learned can be learned from the experience of single members of the human family. Biography, with all due respect, trumps sociology. Piaget based many of his most important contributions on his study of three individuals, his own children. This would strike some contemporary social scientists as virtually unthinkable, but the importance of Piaget's work cannot be denied. It was once a scientific truism that ontogeny recapitulates phylogeny. The story of a single organism's embryonal development is the retelling of the story of a species' evolutionary development. While modern biology has challenged and qualified that truism, it remains the case that in studying the individual we learn about ourselves. In studying the heroic individual we may learn about ourselves in both profound and ultimately very useful ways.

Chapter II

Defining the Hero

> The hero, therefore, is the man or woman who has been able to battle past his personal and local historical limitations.
>
> —Joseph Campbell

What is a hero? Have different ages chosen different types of heroes? Do we choose different heroes at different times of our own lives? What traits does the hero possess? What human needs does the hero serve?

The first English Dictionary (Samuel Johnson's of 1755/1773) defines the hero as: 1) A man eminent for bravery; 2) A man of the highest class in any respect; as, a hero in learning. The *Random House Unabridged Dictionary's* (1993) definitions are strikingly similar:

1) A man of distinguished courage or ability, admired for his brave deeds and noble qualities.
2) A person who, in the opinion of others, has heroic qualities or has performed a heroic act and is regarded as a model or ideal.
3) The principal male character in a story, play, film, etc.
4) A being of godlike prowess and beneficence who often came to be honored as a divinity.

Interestingly enough, *The New Columbia Encyclopedia* (1975) offers an even more narrow description that seems just as out of date:

> Hero: in Greek religion, a famous person, who after his death was worshiped as quasi-divine. The heroes might be actual great men and women, real or imaginary ancestors, or 'faded' gods and goddesses (i.e., ancient gods who for

some reason were demoted to human status). Homer treats his heroes as nobles and fighting men, but many Homeric heroes, such as Hector and Achilles, later became objects of worship . . . .

Recent definitions including those on the internet are similarly classical. The *Merriam-Webster's Collegiate Dictionary*, 10th edition (1993) describes the hero as:

a) a mythological or legendary figure often of divine descent endowed with great strength or ability;
b) an illustrious warrior;
c) a man admired for his achievements and noble qualities; or
d) one that shows great courage.

The American Heritage College Dictionary, 3rd edition (1993) contains a similar definition and also echoes Johnson's eighteenth-century entry:

1) in mythology and legend, a man, often of divine ancestry, who is endowed with great courage and strength, celebrated for bold exploits, and favored by the gods.
2) a person noted for feats of courage or nobility of purpose, esp. one who has risked or sacrificed his or her life.
3) a person noted for special achievement in a particular field: *the heroes of medicine.*

The consistency here is striking. While courage is often considered a characteristic of the hero, the association of the hero with courageous actions is here considered *primary*, as is the association of the hero with military endeavors, military endeavors as represented and imagined in antiquity and associated especially with the writings of Homer and Virgil. In the twenty-first century we live at a considerable distance from many of the notions and ideals of antiquity and our life experiences are quite different from those of ancient Athens, Troy and Rome, even though there are common themes in all human lives and common values across human history. In the present case, the scholarly literature that deals with the subject of heroism is very broad, far more broad than the dictionary definitions of the term would even begin to suggest. In the pages that follow we will be considering a number of individuals whose experience included military activity, but any discussion of heroism that one might hold with a group of contemporary individuals would quickly reveal the narrowness of these definitions. Johnson's second definition ("a man of the highest class in any respect") and Webster's third ("any person admired for his qualities or achievements and regarded as an ideal or model") are far closer to the mark.

In point of fact, different periods of human history have celebrated heroes from very different walks of life (see Figure 1). The pre-Christian era often focused upon demi-gods and warriors, but those of that time also celebrated such figures as philosophers and athletes. Aristotle and others, for example, speak of one of the famous heroes of antiquity, the wrestler, Milo (4th century B.C.).

In the middle ages, martyrs, churchmen, and religious leaders of various kinds were given greater prominence than they would, of course, have received in antiquity and more than they received in the renaissance, which celebrated artists, scientists, and explorers, though the reformation and counter reformation spawned their own succession of martyrs. The scientists and, to a lesser extent, the explorers, were prime heroes of the enlightenment, though there were noteworthy military figures as well.

In the renaissance scientists were sometimes associated with the practice of "black" magic and as a result they were frequently imprisoned and even executed. By the eighteenth century, however, great scientists were generally considered heroes, both because of their ability to provide seemingly comprehensive explanations of the natural world and because of the fact that the skepticism implicit in their methodology had served as an effective intellectual tool to challenge the dominance of the church and the aristocracy in the generations preceding. One thinks, for example, of Pope's lines, "Nature and Nature's laws lay hid in night; / God said, Let Newton be! and all was light."

That skepticism associated with the scientific method cut both ways, however. While celebrated as an intellectual model (Popkin's "constructive skepticism") it could also be used as an effective weapon. In the eighteenth century one can already see a growing skepticism with regard to certain perceived forms of heroism, a skepticism which accompanied the rise of the popular press in the seventeenth century, a skepticism which has dramatically increased in our own time as print journalism has been joined by electronic media to generate facts and information that can both create heroes and deflate them.

We continue to celebrate heroism but we approach depictions of it (particularly the "official" depictions of it) with a great deal of wariness and suspicion. We also still honor military figures, but—with the nineteenth century—we occasionally honor industrial figures and we continue to praise scientists and athletes. We continue to honor artists, with a growing interest in musicians and songwriters/composers and we praise athletes (though with some guardedness in an era of performance-enhancing drugs).

Exploration as an activity has been merged with aviation (a particular area of heroism for women) as the test pilots and scientists who became astronauts have been represented as explorers as well as fliers. We also now honor

Figure 1. Selected Areas of Heroic Endeavor by Era.

CLASSICAL ANTIQUITY	MIDDLE AGES	RENAISSANCE	ENLIGHTENMENT	19th CENTURY	20th CENTURY to PRESENT
Athletes	Clergymen	Architects	Explorers	Captains of Industry	Athletes
Gods	Martyrs	Explorers	Military Leaders	Explorers	Civil Rights Leaders
Mythic Figures	Philosophers	Painters	Philosophers	Inventors	Explorers
Philosophers	Theologians	Philosophers	Religious Leaders	Military Figures	Inventors
Warriors		Religious Leaders	Scientists	Religious Leaders	Investigative Journalists
Writers		Scientists	Statesmen	Writers	Martyrs
		Seamen	Writers		Military Figures
		Settlers			Musicians
		Writers			Religious Leaders
					Scientists
					Statesmen
					Writers

political activists and reformers, missionaries, peacemakers, and a host of individuals whose experience transcends the narrow boundaries circumscribed by earlier (and some current, dictionary) definitions.

Thus, in attempting to formulate a workable conception of the hero, one must move beyond standard definitions. In doing so it quickly becomes apparent that the heroes of one generation are different from those of another, just as the heroes of one individual are not necessarily the heroes of another. There is a great deal of subjectivism in the process and one cannot assume that *any* individual is automatically perceived to be a hero by all.

For example, informal surveys seeking opinions as to who is and who is not heroic will often elicit the name of Dr. Martin Luther King, Jr., but Dr. King has been criticized for the nature of his use of scholarly materials in his research, a matter of great importance to some individuals and of little or no importance to others. For the former the revelations concerning his scholarship invalidated Dr. King as a candidate for heroism; for the latter his use of previous scholarly sources was largely an irrelevance, since his heroism was associated with his political and moral leadership, not his scholarship. The same can be said of revelations/allegations concerning his private life.

For some individuals the private lives of public individuals are of considerable importance; for others they are irrelevant, as the ongoing dialogue concerning several recent presidents demonstrates. Similarly, an individual's public postures can be divided. For some, Lyndon Johnson is the Johnson of the Great Society programs, for others the Johnson of Vietnam. For some, Richard Nixon is the Nixon of the China initiative, for others the Nixon of Watergate.

These are complex matters and one cannot automatically divide individual responses based on the political orientation of those rendering an opinion. Clearly, each of us will tend to focus on different aspects of different potential heroes, but the lives of those potential heroes are of a piece and the fairest response would be to consider the totality of those individuals' lives. That does not mean that we will not still cling to our personal conceptions of the heroic, but it points up the fact that if we truly wish to be fair we should strive to keep in mind our own subjectivism.

That subjectivism remains extremely important, however. One name, for example, that one might expect to elicit some degree of unanimity if one were seeking candidates to categorize as heroes or heroines is that of Mother Teresa. However, we have had students in our classes challenge Mother Teresa as a heroic model on the grounds that she did not permit her associates or subordinates any portion of the limelight in which she appeared. "Can anyone here name Mother Teresa's second-in-command?" a student once asked. Another replied that Mother Teresa had to maintain a position of personal prominence since she was, perforce, a fundraiser and the association of her personality with her cause facilitated the raising of money. To that, another student promptly responded that if she was a fundraiser she could not possibly be heroic, since successful fundraising always taints in some way. Another student responded that every *public* figure who deals with large sums of money is automatically a *political* figure and every political figure is, *a priori*, unheroic.

The influence of the media is strong and its work is in part responsible for the skepticism and even cynicism which characterizes the responses of many contemporary individuals who seek (in vain, they sometimes say) for modern heroes. We now know too much, it is argued. Each individual has strengths and weaknesses, but the weaknesses are so attractive to the tabloid press and electronic media that it is impossible (it is argued) to see individuals with the innocent eyes of past eras. That, of course, suggests that there have never been real heroes, only deluded hero worshippers, but the need for heroes (as we will argue later) continues, even in these jaded days. One could argue that our skepticism has served to increase our need for authentic heroes, since those who might have been easy candidates for heroism in former times are now subjected to criticism that is sufficiently searching to invalidate their candidacy.

Thus, to approach this issue with due attention to its complexity one must consider the subjective elements which play so important a role in its understanding. Theories may be adduced and definitions offered, but, as we have seen, published definitions are quite out of date *vis à vis* contemporary experience and abstract theories must take into account concrete examples and the subjective responses that will inevitably attend those examples.

How are we to establish a path through this thicket? In addition to our readings in the literature of heroism we have been aided immeasurably by our interactions with individuals with strong and often suggestive views. Much of the material in this chapter grows out of those interactions and the charts and figures which we include have been continually refined by our discussions with a spectrum of students ranging from young adults to senior citizens. We have had students who have fought in wars and those who have only seen wars on television (or played them on video games). They have come from all parts of the country and the world, from all walks of life and with all levels of education. They have told us of their heroes and, in a number of cases, argued that the very notion of the hero is unthinkable.

While the issue has many subjective dimensions, as previously noted, it can still be discussed and described with a fair degree of objectivity, since key notions and themes continue to repeat in every discussion of the subject. In the next chapter we will discuss the relationship between heroism and human development, but it is useful to begin here with the observation that one's heroes are to a considerable degree a function of one's personal time line. Each of us enters human history at a different point and the events which mark our history can be of greater or less importance to us depending on the confluence between our own developmental processes and the historical patterns which surround us. (It has been said, for example, that our personal political stance and general view of the world is based on the state of that world at the time when we were teenagers.)

To take a simple example, on February 3, 1959 Buddy Holly, Ritchie Valens, and J. P. Richardson (the "Big Bopper") died in a post-concert plane crash outside of Clear Lake, Iowa. In a song entitled "American Pie" (1971)—the name once rumored to be the name of Holly's plane—Don Maclean memorialized this event by describing it as "the day the music died." This was a defining event for Maclean's generation, the generation, for example, of the Beatles, who had named their own band in homage to Buddy Holly's band, the Crickets.

For those in their mid and late teens in 1959 (such as the members of the Beatles) this was an extremely important event. However, it is not enough that one be alive when events such as this occur. It is also important that there be some relationship between one's own life and those events. For an individual who was three years old or sixty-five years old in 1959 the event was, in all likelihood, far less important. For an individual who was a teenager during those years, the years in which rock music became not only established but for many all-important, this event was shattering. As a young professor I (RBS) witnessed a cognate example of this phenomenon. Walking down a dark hallway of an academic building one day I saw a frail, public announcement scrawled on a portable blackboard: *Janis Joplin died today*. I remember the event vividly. I also remember my older colleagues walking past it, asking who Janis Joplin was. More recent deaths (Kurt Cobain, Tupac Shakur, e.g.) evoke similar responses (strong, weak) depending on the personal timelines of those reacting to them.

Hence, to understand our own views of heroism, it is useful to construct an historical time line of the twentieth and twenty-first centuries and then place our personal time line against that larger pattern. Realizing the importance of our own developmental processes in our personal judgments of that which is or is not heroic serves to make us aware of our own subjectivism and demonstrates one of the reasons why there are disagreements between individuals concerning heroism even though the people and events in question seem to *us* to be unquestionably heroic (or *unquestionably* unheroic).

These disjunctions are striking and they are exacerbated by the fact that the generations living today have seen so much change in such a short time.

Moreover, they have had those changes highlighted by and intensified by the media. While the great depression may be a defining event for our parents or grandparents it is of little consequence to those young people for whom it represents a chapter in a history book rather than an event of compelling personal significance. The same is true of such events as the Japanese attack on Pearl Harbor, the detonation of the atomic bomb, the rise of the Third Reich, the holocaust, and, increasingly, the civil rights movement, Vietnam, and such events as Watergate or Chappaquiddick.

An historical time line is useful in elucidating the manner in which a given period will conceptualize its heroes. The twentieth century, for example, was marked by a succession of world wars and then so-called brushfire wars. It is a century of assassinations and attempted assassinations, a century of political movements involving one form or other of liberation, a century of scientific/technological breakthroughs and ecological disasters. Most of these events have been precipitated by human actions and they have in turn precipitated human responses, actions and responses which have altered our conception of heroism, particularly to the degree to which we have been personally affected by the events.

Standing at a historical site one experiences these matters directly. In the Cabinet War Rooms in London, for example, one sees young students from all over the world responding to the museum as one responds to distant history; at the same time one routinely sees older British citizens with tears streaming down their cheeks as they relive a part of their early lives.

On the other hand there are those who have been able to personalize their country's history, no matter how distant that history might be from current experience. Tears are still shed by Scottish men and women on Culloden moor, even though the defeat of the forces (of, to some, *Bonnie Prince Charlie*, to others, the *Pretender*) there occurred in 1746 and at night on Little Round Top one commonly finds young people attempting to feel what the federal troops might have felt before the second day's battle at Gettysburg (just as groups of people of all ages walk the mile between Seminary Ridge and Cemetery Ridge in rough formation, trying to feel what the confederate troops might have felt in Pickett's Charge).

The twentieth century was witness to a succession of important events, too numerous to mention individually. We have included a time line (see Figure 2) which includes representative occurrences that have affected our consciousness and our sense of the heroic, a sense based both on events and on cultural representations. We invite the reader to sketch in his or her personal time line and consider the confluence between important events in his or her own life and those in the life of our society. This is an excellent entrée into the subject for those wishing to discuss heroes and heroism with other people.

Figure 2. Personal/Historic Timeline [selected events]

To position your life within the context of world events, place your own dates on the time line.
List positive world events above the line, negative events beneath it. A selected set of events is listed for your consideration.

Birthdate ______________________________ 2009

1898 - Discovery of Radium (Madame Curie)
1899 - Boer War [1899-1902]
1902 - First Movie Theater Opens
1904 - First U.S. Olympic Games (St. Louis)
1906 - San Francisco Earthquake
1907 - Albert Einstein proposes the Theory of Relativity
1907 - Picasso's *Les Demoiselles d'Avignon* introduces cubism
1908 - Assembly Line manufacturing created
1908 - Model-T Invented
1909 - First Airplane Flight (Wright Brothers)
1909 - Roald Amundsen reaches North Pole
1912 - *Titanic* Sinks
1913 - Igor Stravinsky's *The Rite of Spring* first performed
1913 - U.S. Income Tax Mandated
1913 - Zipper Invented
1914 - WW I [1914-1918]
1917 - Russian Revolution
1920 - League of Nations Established
1920 - Prohibition [1920-1933]
1921 - First Radio Broadcast
1922 - Insulin Discovered
1922 - James Joyce's *Ulysses* published
1923 - King Tut's Tomb Opened
1924 - Citizenship Granted to Native Americans born in USA

1925 - Adolf Hitler's *Mein Kampf* published
1925 - F. Scott Fitzgerald's *The Great Gatsby* published
1925 - Scopes Monkey Trial
1927 - Monsignor Georges Lemaître proposes big bang theory
1927 - Philo Taylor Farnsworth demonstrates working model of a television
1928 - Sir Alexander Fleming discovers Penicillin
1929 - Edwin Hubble proposes theory of expanding universe
1929 - First Talking Movie
1929 - Great Depression [U.S. 1929-WW II]
1929 - Lindbergh's Trans-Atlantic Flight
1929 - Vatican City becomes an independent state
1930 - Pluto Discovered
1931 - Empire State Building Completed
1932 - Iraq becomes an independent country
1933 - Adolf Hitler becomes Chancellor of Germany
1933 - Wiley Post's Solo Flight Around the World
1936 - John Maynard Keynes's *The General Theory of Employment, Interest, and Money* published
1936 - Margaret Mitchell's *Gone With the Wind* published (Awarded the Pulitzer Prize for Literature in 1937)
1937 - First Blood Bank Established
1937 - Hindenburg Disaster
1939 - First Football Game TV Broadcast
1939 - John Steinbeck's *The Grapes of Wrath* published (Awarded the Pulitzer Prize for Literature in 1940)
1939 - WW II [1939-1945]
1940 - Ernest Hemingway's *For Whom the Bell Tolls* published
1940 - Winston Churchill becomes Prime Minister of Great Britain
1941 - Japan Bombs Pearl Harbor
1941 - The Manhattan Project
1941 - U.S. declares war on Japan
1944 - Allied D-Day Invasion
1945 - Allied Victory in Europe
1945 - Bombing of Hiroshima and Nagasaki

(*continued*)

Figure 2. ***(continued)***

1945 - Allied Victory in Japan
1945 - First electronic computer, ENIAC, is built
1945 - Microwave oven patented
1945 - Yalta Peace Conference
1947 - Dead Sea Scrolls found
1947 - Jackie Robinson (First Modern Era African-American Baseball Player)
1947 - Tennessee Williams' *A Streetcar Named Desire* first performed (Awarded the Pulitzer Prize for Drama in 1948)
1948 - Apartheid formalized
1948 - Berlin Airlift
1949 - George Orwell's *1984* published
1950 - Korean War [1950-1953]
1950 - McCarthyism [1950-1957+/-]
1951 - Adolf Eichmann Trial
1951 - Julius and Ethel Rosenberg Trial
1952 - J.D. Salinger's *A Catcher in the Rye* published
1953 - Elizabeth II of Great Britain crowned
1953 - Mt. Everest summited by Hillary and Norgay
1953 - Watson, Crick, and Franklin discover DNA's structure
1954 - Brown v. Board of Education begins unraveling of U.S. racial segregation
1954 - J.R.R. Tolkien's *The Lord of the Rings* published
1956 - Hungarian Uprising
1956 - Oral Polio Vaccine invented
1957 - Jack Kerouac's *On the Road* published
1957 - Russia launches first satellite, *Sputnik I*
1957 - Satellite Communication
1958 - *USS Nautilus* (SSN-571) First watercraft to reach the geographic North Pole
1959 - Death of Buddy Holly
1959 - Mary and Louis Leakey uncover hominid fossils
1959 - Vietnam War [1959-1975]
1960 - Beatles Phenomenon [1960-1970]

1960 - Harper Lee's *To Kill A Mockingbird* published (Awarded the Pulitzer Prize for Literature in 1961)
1961 - Birth Control Pill approved by FDA
1961 - Civil Rights March in Selma, Alabama
1961 - Construction of Berlin Wall
1961 - First Man in Space (Soviet Cosmonaut, Yuri Gagarin)
1961 - Joseph Heller's *Catch-22* published
1962 - Cuban Missile Crisis
1962 - John Glenn (Piloted the first American manned orbital mission)
1963 - Betty Friedan's *The Feminine Mystique* published
1963 - John F. Kennedy Assassinated
1965 - First Woman in Space (Valentina Tereshkova from the Soviet Union)
1967 - First Human Heart Transplant performed by Dr. Christiaan Barnard
1968 - Martin Luther King, Jr. Assassinated
1968 - Robert F. Kennedy Assassinated
1969 - First Lunar Landing (United States)
1969 - U.S. Astronauts Armstrong and Aldrin walk on the Moon
1970 - Apollo 13 returns safely
1970 - CD-ROM invented (James T. Russell)
1970 - Personal Computer Introduced
1971 - First World Space Station
1972 - Munich Olympic Games
1972 - Watergate
1973 - Mobile Phones Invented
1974 - Patty Hearst Kidnapping
1974 - President Richard Nixon resigns
1975 - Pol Pot Massacre
1975 - U.S. and Soviet spacecraft dock in space
1976 - Alex Haley's *Roots* published
1976 - Episcopal Church ordains women
1976 - First successful landing on Mars (*Viking 1*)
1977 - Death of Elvis Presley
1977 - First *Star Wars* Film

(*continued*)

Figure 2. *(continued)*

1979 - Margaret Thatcher becomes first woman Prime Minister of Great Britain
1980 - Mount St. Helens Erupts
1980 - Smallpox eradicated
1981 - AIDS Pandemic [First Reported, 1981]
1983 - First American woman in space (Sally Ride)
1984 - Ethiopian Famine
1986 - Chernobyl nuclear plant accident
1986 - Space Shuttle Challenger Disaster
1988 - Pan Am Terrorist Attack (Lockerbie, Scotland)
1988 - Stephen Hawking's *A Brief History of Time* published
1989 - Exxon Valdez Disaster
1989 - Fall of Berlin Wall
1989 - Fall of Communism in Eastern Europe
1989 - Globe Theatre reconstructed
1989 - World Wide Web Invented (Tim Berners-Lee)
1990 - Apartheid Dismantled
1990 - Operation Desert Storm
1991 - Dissolution of USSR
1992 - Bosnian War
1993 - Beanie Babies introduced
1993 - Waco Disaster
1993 - Whitewatergate
1995 - O.J. Simpson Trial
1995 - Oklahoma City Bombing
1996 - Kosovo War
1997 - Death of Princess Diana
1997 - Inaugural Harry Potter Book published: *Harry Potter and the Philosopher's Stone*
1998 - Clinton/Lewinsky Scandal
1998 - President Clinton's Impeachment
1998 - Titanic Found
2000 - Dot-com Bubble Burst

2000 - U.S. Presidential Election Contested
2000 - USS Cole Bombing
2001 - Terrorist Attacks in United States
2002 - Rick Warren's *The Purpose Driven Life* published
2002 - Space Shuttle *Columbia* Disaster
2003 - Dan Brown's *The Da Vinci Code* published
2003 - Operation Iraqi Freedom
2003 - President Bush signs $350 billion tax-cut bill
2004 - Facebook website launched
2004 - South East Asia Tsunami
2005 - Hurricane Katrina disaster
2005 - YouTube Invented
2006 - Liquid water is discovered on Enceladus, the sixth largest moon of Saturn
2006 - North Korea launched at least two short-range Nodong-2 missiles, one SCUD missile and one long-range Taepodong-2 missile
2006 - Saddam Hussein is executed by hanging
2006 - The U.S. Census Bureau estimates that the population of the United States has reached 300 million
2007 - First Woman Speaker of the US House of Representatives
2007 - Apple Inc CEO, Steve Jobs announces the iPhone
2007 - *Harry Potter and the Deathly Hallows*, the final book in the bestselling Harry Potter series is released
2007 - The tomb of Herod the Great is discovered
2007 - Virginia Tech massacre: The deadliest mass shooting in modern American history
2008 - Fidel Castro resigns as President of Cuba
2008 - "Bloody Friday" saw many of the world's stock exchanges experience the worst declines in their history
2008 - First day of legal same-sex marriage in California
2008 - Two skeletal remains found near Ekaterinburg, Russia were confirmed to be the remains of Alexei Nikolaevich and one of his sisters
2009 - Inauguration of First African-American U.S. President - Barack Obama
2009 - U.S. General Motors files for bankruptcy
2009 - U.S. Airways Flight 1549 makes emergency landing in the Hudson River. All passengers and crew survive.
2009 - U.S. Navy rescues captain Richard Phillips, killing three pirates and capturing a fourth
2009 - U.S. President Barack Obama loosens restrictions on travel to Cuba
2009 - Michael Jackson "King of Pop" dies
2009 - Sonia Sotomayor confirmed to the U.S. Supreme Court. Sotomayor is the first Hispanic Supreme Court justice.
2009 - Rio de Janeiro wins bid for the 2016 Olympics. It will be the first South American city to host the Games.

Having measured ourselves against a larger historical time line, it is also useful to consider ourselves and our heroes within a set of broad categories (see Figure 3), contextually progressing from personal spheres to the public arena. Each of us leads a personal life; each of us is part of one or more nations, each of us is a citizen of the world, and each of us is a participant—to a greater or lesser extent—in human history. We also see and categorize our heroes in these terms.

Each of us, one would hope, should have heroes who have touched us personally and directly. These would be individuals who are not known to the wider world but who are, nonetheless, heroes to us. This would include such individuals as parents, siblings, relatives, friends, teachers, nurses, and coaches—the sort of individuals who are frequently conceptualized as "unsung" heroes, people who have touched our lives and inspired us in ways that the general public will never fully know and appreciate.

At the same time, there are heroes who are peculiar to individual nations. In the United States one thinks of military figures such as Audie Murphy or Sergeant Alvin York, athletes like Jim Thorpe, Wilma Rudolph or the one-armed pitcher, Jim Abbott. One thinks of political activists such as Rosa Parks or César Chávez, an aviator such as Amelia Earhart or a courageous photographer like Margaret Bourke-White. Many would see individuals like these as heroes, but they are U.S. heroes, not, in general, global heroes.

Global heroes might include such individuals as Jacques Yves Cousteau or Nelson Mandela, Pope John XXIII, or world-renowned athletes such as David Beckham or Tiger Woods. The accomplishments and/or the recognition of these individuals transcends national borders (and—in the case of the athletes—they participated in specific sports which are played globally). At the same time, there are individuals whose accomplishments are now a part of common human history. One thinks of such people as Joan of Arc, Michelangelo, Isaac Newton, Albert Einstein, Galileo, Gandhi, Aristotle, Copernicus, Madame Curie, Martin Luther King, Jr., Albert Schweitzer and Anne Frank. These individuals will not be heroes to all, but the stage on which they are now judged is a historical one.

There is a developmental dimension here to which one should call attention. Some heroes exert a force which grows with time. Beginning as national heroes their work takes on global proportions and then passes into history. Gandhi is a good example here, as is Churchill. At the same time, some of these individuals are more or less successful at the "ordinary," personal level. One wishes, for example, that "unsung" personal heroes could receive the recognition that they truly deserve but seldom receive. At the same time, one would hope that those individuals who are honored by their nations and

Figure 3. Four Categories of Proposed Heroes.*

ORDINARY HEROES	NATIONAL HEROES	GLOBAL HEROES	HISTORICAL HEROES	HISTORICAL HEROES (Cont.)
Ancestors	Henry Aaron	Marian Anderson	Dante Aligheri	Meriwether Lewis
Children	Susan B. Anthony	Corazon Aquino	Roald Amundsen	Abraham Lincoln
Clerics	Margaret Bourke-White	Neil Armstrong	Aristotle	Martin Luther
Coaches	César Chávez	Arthur Ashe	Jane Austen**	Beryl Markham
Mentors	Crazy Horse	Christiaan Barnard	Winston S. Churchill**	Michelangelo
Nurses	Medgar Evers	David Beckham	William Clark	Mother Teresa
Parents	Henry Ford	Tim Berners-Lee	Copernicus	Isaac Newton
Physicians	Yuri Gagarin	Rachel Carson	Marie Curie	Florence Nightingale
Relatives	Lou Gehrig	Jacques Yves Cousteau	Charles Darwin	Tenzing Norgay
Siblings	John Glenn	Michael DeBakey	Albert Einstein	Jesse Owens
Teachers	Billy Graham	Bob Dylan	Enrico Fermi	General George S. Patton, Jr.**
	Chief Joseph	Bill Gates	Anne Frank	Linus Pauling
	General Robert E. Lee	Bob Geldof	Galileo Galilei	Plato
	Christa McAuliffe	Dag Hammarskjöld	Mohandas Gandhi	Ronald Reagan
	Audie Murphy	Stephen Hawking	Edmund Hillary	Jackie Robinson
	Vice-Admiral Horatio Nelson	Mairead Corrigan Maguire	Ignatius of Loyola	Franklin Delano Roosevelt
	Sandra Day O'Connor	Nelson Mandela	Thomas Jefferson	Theodore Roosevelt
	Rosa Parks	Pelé	Joan of Arc	Andrei Sakharov
	Wilma Rudolph	Jonas Salk	Pope John XXIII	Albert Schweitzer
	Vice-Admiral James Stockdale	Margaret Thatcher	Pope John Paul II	Ernest Shackleton
	Jim Thorpe	Desmond Tutu	Samuel Johnson	Socrates
	Ted Williams	Lech Walesa	Jan Karski	Corrie ten Boom
	Frank Lloyd Wright	Elie Wiesel	Helen Keller	Raoul Wallenberg
	Sergeant Alvin C. York	Tiger Woods	Martin Luther King, Jr.	George Washington
			T. E. Lawrence**	Woodrow Wilson
				Orville and Wilbur Wright

*Proposed by our students; see Appendix for further information.

**These individuals are discussed in the book.

(Cont. Next Column)

whose public works are enshrined in human history were also capable of serving as heroes to those who knew them best.

There is an old maxim (variously attributed to Montaigne, Madame de Sévigné and Madame Cornuel) that no man can be a hero to his own valet (just as no man can be a prophet in his own village), but this notion is regularly contradicted by experience. Many of the heroes who have achieved global acclaim also enjoyed the deep affection and respect of those closest to them. Homer, to return to our first example, clearly intends this to be the case with Odysseus. Familiarity need not breed contempt and it need not breed cynicism. Proximity can in fact lead to increased admiration and inspiration, for those closest to the hero or heroine are aware of a host of daily trials and triumphs which the public never knows.

At the same time, we must remember the importance of timing and luck. In what may be the most frequently quoted poem in English, Thomas Gray's "Elegy Written in a Country Churchyard" (1751), the poem's speaker points out that the now anonymous men and women buried in a rural churchyard might have included individuals who could have attained public, heroic stature if history had only cooperated:

> Perhaps in this neglected spot is laid
> Some heart once pregnant with celestial fire;
> Hands that the rod of empire might have swayed,
> Or waked to ecstasy the living lyre.
>
> But Knowledge to their eyes her ample page,
> Rich with the spoils of time, did ne'er unroll;
> Chill penury repressed their noble rage,
> And froze the genial current of the soul.
>
> Full many a gem of purest ray serene,
> The dark unfathomed caves of ocean bear;
> Full many a flower is born to blush unseen,
> And waste its sweetness on the desert air.
>
> (ll. 45–56)

Passages and apothegms continue to be cited and quoted because they carry some portion of the truth for a significant number of people. In this case, it is clear that we believe both things: it is not easy to be a hero—either to the multitudes or to those who know us intimately. At the same time, it is clear that the nobility and heroism of a great number of individuals goes unrecognized in the flow of human history. The world is not always "fair."

Time does not always cooperate, but it would seem fair and fitting to many of us if many heroic individuals now forgotten had not been forgotten, just as it would prove particularly satisfying to learn that those individuals idolized by mankind were similarly loved and admired by those who knew them on a personal basis. The interplay between the public and the private will be one of our continuing concerns as we proceed.

Heroes have been "found" in many areas of human endeavor: in the military, in athletics, in government and business, in the arts, in the world of political activism, in medicine, science, and the world of intellect, in aviation, in religion, and in the world of philanthropy, to name but a few. However, there is still a great deal of room for subjectivity in this list. Some have argued that the profit motive of modern business forces any possibility for true heroism to the margins of the enterprise. Similarly, some have argued that modern science is now so thoroughly institutionalized and so dependent upon significant financial support (often from the federal establishment), that the lone inventor or crusading individual is largely a fixture of the past. It is possible, of course, to move between these areas. Businessmen become philanthropists. Musicians and actors become politicians. Robber barons become public benefactors. Scientists and artists become political activists, just as former political activists sometimes become businessmen.

In any discussion of specific heroes or specific areas of potential heroic behavior there will be disagreements as to whole categories and classes. There is *less* disagreement with regard to the *characteristics* which mark the hero. Individuals will dispute whether this or that person actually possesses some or all of these characteristics, but there is usually general consensus with regard to the characteristics themselves. The lineaments of the hero are not, however, absolute. The realization that our views of heroism (and of particular heroes/heroines) shift over time has been reaffirmed recently by Stephanie Barczewski. In her study *Antarctic Destinies: Scott, Shackleton and the Changing Face of Heroism,* she charts the manner in which history's judgments of Scott and Shackleton have materially changed. In his own time Scott often overshadowed Shackleton, but that judgment has shifted and Shackleton's accomplishments have now moved to the foreground.

In our discussions with a wide array of participants, the following characteristics of the hero or heroine have generally been highlighted (see Figure 4).

Obviously, not every hero will possess each of these characteristics and different observers will place different weight on different traits and features. In his dialogue the *Gorgias*, for example, Plato places great emphasis on discipline:

Figure 4. Heroic Characteristics.

- The hero is a model.
- The hero is selfless rather than selfish.
- The hero is hopeful.
- The hero acts in order to do the right thing.
- The hero is a risk-taker.
- The hero seeks to make a difference.
- The hero is an idealist.
- The hero is a more real and authentic person than the average individual.
- The hero is focused.
- The hero has willpower.
- The hero has competence.
- The hero has judgment and discernment.
- The hero is disciplined.
- The hero's actions are remembered, though the hero does not act simply in order to be remembered.
- While the hero may act out of anger or a sense of injustice, s/he acts to achieve positive results.
- The hero is able to love.
- The hero is fortunate because circumstances cooperate with her/him.

> The man who is disciplined will behave with propriety towards God and man; if he behaved improperly he would not deserve the name of disciplined. That is undeniable. Again, proper behavior toward men is uprightness and proper behavior toward God reverence; and a man who acts uprightly and reverently must be upright and reverent. Certainly. And not only upright and reverent but brave as well; a disciplined man will not choose inappropriate objects either to pursue or to shun; on the contrary he will pursue or shun the things and people and pleasures and pains that deserve either course, and he will stand his ground firmly where duty requires it.
>
> (p. 116)

The first characteristic (to which we will return later) is especially important. The hero is a model. We speak of heroic actions but we are generally hesitant to consider an individual a hero unless we know the context of those actions. The context may consist of many things, but it certainly includes the arc of the individual's life and the values and experiences which have shaped that life.

The 1994 Robert Zemeckis film, *Forrest Gump* (based on an earlier novel), depicts an individual of below-average intelligence who participates in many of the great events of his time and wins such awards as the Congressional Medal of Honor. He wins this medal for carrying his comrades away from a battle zone that is under fire. He is remembered; he is selfless; he is seeking positive results; he is hopeful; he is attempting to do the right thing; he is taking risks, seeking to make a difference, and embodying idealism. In his way he is focused; he is lucky; he possesses willpower, physical competence, and the ability to love. He also has discipline, though his judgment and discernment need to be balanced with luck.

However, one shift in the pattern of his life and outlook could invalidate his claims as a hero. Saving one's comrades may be considered, *a priori*, a heroic act, but we would not consider it heroic if we knew that the person involved was motivated solely by a desire for self-aggrandizement. We would not consider it heroic if the individual performed the act in hopes of capitalizing on his experience financially and we would judge it less heroic if the individual were somehow physically impaired and was not actually aware of the physical dangers surrounding him.

In short, we want to see the heroic act or acts as part of a pattern of behavior and thus we characteristically do describe such behavior contextually. An illustrative example of the importance of the pattern is the series of vignettes which the sports divisions of the networks shoot in which Olympic stars are seen, as they say, "up close and personal." It is not enough to see them compete in their particular event or events. The networks show what the audience wants to see: the context of the individual athletes' experience—their personal lives and personal stories, their personal challenges and personal tragedies. These stories (particularly the "special" ones) help the audience to contextualize the athlete's physical accomplishments within a larger pattern of experience, one which adds meaning and poignancy to a simple physical event.

It is not enough to see Dan Jansen skate for a gold medal. We must also know the extent to which he has failed to achieve such a medal in the past. We must see his physical gifts in the context of personal hope, personal striving, and personal tenacity. We also see that effort within the context of his family, with his loved ones suffering and triumphing with him and his young daughter, Jane—named for his deceased sister—being carried in his loving arms on his victory lap.

When the context disappoints us we are capable of altering our view of the potentially heroic action. Nancy Kerrigan, for example, skated after being subjected to a violent physical attack. This backdrop was given more attention than her actual athletic performance and many saw her, sympathetically, as a woman who nearly missed the chance of a lifetime to compete and then bravely prepared herself for the event and gave one of her finest performances in the face of physical pain and overexposure from the media. However, after the event, she was criticized harshly for comments made while awaiting receipt of her medal and further comments made during a personal appearance at a Disney theme park. The actions (in this case, participation in the Olympics) in themselves are insufficient; we always wish to see the larger experiential and behavioral context before we decide upon our heroes and heroines.

One reason for this, of course, is that not all of us can imitate specific actions. We cannot all be astronauts. We cannot all be born in times which are propitious for our particular abilities and virtues; certain things are already discovered; certain devices have already been invented. We can, however, be inspired by a pattern of activity. For example, Forrest Gump loves even when his love is not reciprocated. That is a part of the pattern of his experience and behavior. It characterizes and contextualizes his potentially heroic acts and makes them (for many) truly heroic.

If there is some consensus with regard to the characteristics of the hero there is similar consensus with regard to the human need for heroes. Heroes inspire us and we all require inspiration. The word, from *in* + *spirare*, literally means to be breathed into, to be filled with spiritual sustenance. We need this spiritual and psychological sustenance just as we need oxygen in the more mundane (but no less crucial) physical aspect of our lives. We cannot live without oxygen; we cannot live *meaningfully* without inspiration. That is why inspiration is often associated with the divine and/or heavenly since our need for it is so great that we link it with the transcendent.

The need for heroes is related to the characteristics of the hero. The hero exemplifies traits and demonstrates capabilities—traits and capabilities which we value. He or she extends our conceptions of the possible and helps us to focus on that which is essential rather than peripheral. The hero helps us to consolidate our thoughts by serving as a living exemplar of our most deeply held beliefs. By his or her very existence the hero speaks to and meets essential human needs. These are summarized in Figure 5.

This need is so strong that the media will manufacture pseudo-heroes in order to meet it. Thus, we see such individuals as Fabio, a man whose principal claims to fame are the length of his hair, the breadth of his chest, and his multiple appearances on the jackets of romance novels. He is a media fabrication. He is not a hero, but rather a pseudo-hero, an individual who looks like a

Figure 5. The Need for Heroes.

- The hero demonstrates that an individual *can* be special (and hence we can too).
- The hero demonstrates how a person can be real (and hence we can too).
- The hero extends the notion of human capability (genius, endurance, etc.).
- The hero shows that we can rise above the ordinary (of which we have a surfeit).
- The hero shows that the old ideals are not dead (rises above selfishness, venality, and bureaucracy).
- The hero shows how we can focus on what is *important* in life.
- The hero can *embody* grand ideas (aspirations, political visions, etc.).
- The hero shows that we can be remembered.

larger-than-life character from fiction. This is similar to what Daniel Boorstin terms the pseudo-event: a manufactured event constructed to give the illusion of reality and authenticity ("the *first, annual* awards show of its kind"). Such individuals and events spark interest in ways that mirror the larger, deeper, more authentic interest sparked by real heroes and real events. They have to suffice when the real thing is not available. The frequency of their occurrence in our own time points to the fact that our need is especially deep and the number of authentic heroes especially (or apparently) small.

Creations of the media are not, as we indicated earlier, un-instructive. The media provide, in part, what the public desires and while the media shape public taste they also respond to it. We can learn a great deal about the manner in which our society imagines heroes and heroism by looking at such characteristic cultural products of our society as popular films.

Raiders of the Lost Ark (1981), Steven Spielberg's immensely successful film, inaugurated the Indiana Jones series, a group of films dealing with the exploits of an adventurous professor of Archaeology. The series was itself based on the Saturday-afternoon serials which Spielberg enjoyed in his youth. In particular, *Raiders of the Lost Ark* was based on the serial, *Manhunt in the African Jungle* (1943), a Rod Cameron vehicle concerning Nazis in north Africa who are attempting to rally support for their cause through the fabrication of a document that is associated with an ancient dagger, an artifact revered in Arabia. The document urges all Arabs to follow the sign of the swastika.

The serial is extremely crude. There are, for example, despite the title, no real manhunts and no scenes in the jungle. An airport which is identified as being in Nazi Germany has palm trees clearly visible in the background; it is obvious that we are in Burbank, not Berlin. Nevertheless, the serial made an impression on Spielberg, an impression that helped to inspire the making of *Raiders of the Lost Ark*, a film which found a very large and very receptive audience.

The hero of *Manhunt in the African Jungle*, like so many of the Saturday serial heroes, is explicitly "American," an adjective which connotes a succession of characteristics. He is honest, forthright, and always fair. He is also strong, clean-cut, and strikingly innocent. He cooperates with his allies (in this case, Duncan Renaldo, playing a Frenchman instead of the Cisco Kid) and has a female sidekick who is also quintessentially "American." She is an investigative reporter who is best described as "spunky," particularly for her time. She is also able to fly airplanes, an activity frequently connected in the public mind with the modern heroine. One thinks of such real life examples as Beryl Markham, Amelia Earhart, and, later, Christa McAuliffe.

The true, brave, steadfast, "American" hero is a mainstay of our popular culture and the range of examples of cinematic heroism is wide indeed. One of the most popular of such heroes is the defender of justice and bulwark against evil, the detective. John Cawelti has argued that the figure of the detective arises in the popular culture of the nineteenth century as a response to a restive proletariat and still powerful, though declining, aristocracy. The detective represents middle class rationalism, a rationalism built upon science and distinguishable from such alternative forces as group movements (the proletariat) or dictatorial authority (the church and the aristocracy). Wilkie Collins' Sergeant Cuff (in *The Moonstone*, 1868) is an early example, but the great example is Arthur Conan Doyle's Sherlock Holmes.

The Holmes legend has spawned endless literary and cinematic reproductions and a host of Holmes societies. The most famous and influential of the Holmes films, however—the Rathbone/Bruce films —so English in their tone and, usually, in their setting, are the products of Hollywood. The fog-drenched moors, wind-swept tors, and alleys and byways of London are all actually California sound stages.

It is appropriate that a society with so broad a middle class as America's would take to the Holmes legend and be attracted to the master detective's use of hard fact, solid evidence, and clear reasoning. Holmes also indicates something of the American's awe of England, particularly in its language, its traditions, and its customs.

The common man is a frequent (and expected) hero in American cinema—from the films of Frank Capra down to such Spielberg films as *Jaws* (1975). However, the master of this theme was Alfred Hitchcock, an English director who began his work in his own country but later achieved great success in Hollywood. Hitchcock's standard plot concerns a common man thrown into a difficult and tense situation who faces that situation and acquits himself well. *North by Northwest* (1959) is a frequent example, as are *Rear Window* (1954), *Vertigo* (1958), *Frenzy* (1972), *The Saboteur* (1942) and *The 39 Steps* (1935).

The 39 Steps is based on a 1915 novel by John Buchan, to whom we will return later in our discussion of T. E. Lawrence. Suffice to say here that Buchan's nomination of Lawrence as a possible hero for Lowell Thomas can be connected with another popular cultural hero, Lester Dent's Doc Savage, one of the original models behind the extremely influential creation of Ian Fleming, James Bond.

Bond is another English creation who has found an enormous following in America. At the time of the initial films (*Dr. No*, *From Russia with Love*) in 1962–3 the world was surprised to hear that many high school and college students had inquired of the British government whether or not positions such as James Bond's might be available for summer work.

Bond is, in part, a techno-hero and his success can be related to the successes of such writers as Tom Clancy, who now flourish in the novelistic genre of the so-called techno-thriller. Where the classic detective utilizes scientific rationalism—a middle-class intellectual instrument—the techno-hero utilizes science as an extension of human physical and mental capability. This is a fitting hero for a society as technically oriented and technically proficient as our own, particularly when we are struggling with the challenges of constant technological change and what Alvin Toffler in 1970 termed "future shock."

One of the most influential of the American heroes is the western hero, a type with a large following both in Europe and in Asia. The western hero is also a common man, but one measured against landscape rather than technology. The close relationship between American landscape and the growth and structuring of American society has long been noted. The so-called Turner hypothesis (after the University of Wisconsin historian, Frederick Jackson Turner) concerns the manner in which the country has been culturally defined by the receding, western horizon.

The western hero suggests the omnipresent conflict in American society between freedom and constraint. America is the only major industrialized country in the world, for example, which does not require its citizens to carry identification cards. Social Security cards specify that they are not to be used for identification purposes. That is because one of our cultural presuppositions is that an individual should be able to redefine himself or herself (though usually it is himself) by simply "setting out for the territory" and leaving his usual environs. The Kevin Costner film, *Dances With Wolves* (1990), is an extreme version of the pattern. The protagonist not only sets out for the frontier; he becomes, in a sense, a native American.

One film which neatly encapsulates these ideas is *The Man Who Shot Liberty Valance* (1962). The story concerns a lawyer who attempts to practice law in the new but still wild west. He confronts lawlessness in the form of a ruthless criminal (an "outlaw") who is fond of not only plying his trade but also destroying such instruments of eastern civilization as the local newspaper's printing press. The lawyer faces the outlaw in a shootout, but the outlaw is actually killed by another man standing in the shadows, protecting the lawyer, whose chances against the outlaw would have been nil. The protector is himself a free spirit, uncomfortable with the incursions of civilization, but also a fundamentally good, fair man. The film demonstrates the importance, in America, of the border between freedom and the law, as well as the border between the frontier and the city and notes that while we must change—we cannot remain eternally on the frontier—that fact is also a great pity.

The interesting thing about *The Man Who Shot Liberty Valance* is that the roles are all played by cultural icons. Thus, the message of the picture is conveyed by the players (each of whom radiates cultural associations) as well as by the script. The lawyer is played by Jimmy Stewart, the outlaw by Lee Marvin, and the good, fair, free spirit in the shadows by John Wayne.

This illustrates the manner in which we conflate our notions of heroism with cultural representations. Clark Gable, for example, is inevitably associated with romantic lead roles, while John Wayne is associated with western and military roles. The fact that Clark Gable was actually in the military (as was Jimmy Stewart, an Air Force general) and John Wayne never served is often overlooked. The representations and the cultural associations are stronger than the realities of the situation.

Along with our heroes we have also developed more than our share of antiheroes. The noir hero (as in *Chinatown* [1974] or *Dirty Harry* [1971]) comes out of the Raymond Chandler tradition of hardboiled detective fiction, one in which a stark and realistic figure confronts a stark and realistic setting, a setting in which rationalism is not always effective. The British detective (i.e. the detective in the traditional English mystery story) triumphs with his or her brains and restores order to a society that has been temporarily disrupted. The noir hero, on the other hand, attempts to survive in a corrupt society where large institutions (political, criminal, industrial, governmental) frequently triumph over the lone individual. The Chandler hero is uncommonly effective in this setting, but his cultural resonance results from the fact that he also represents the common man.

It is fascinating to trace the origins of such cultural constructs. In his recent biography of Chandler (1997), Tom Hiney has pointed out that Chandler and several of his fellow graduates of Dulwich Academy—Ernest Shackleton's alma mater, near Greenwich—were all influenced greatly by the headmaster,

A. H. Gilkes. Gilkes had no patience with arrogance or pretense and he was a great defender of the common man against elitism of any sort. Hiney speculates that it may well be the influence and outlook of this great teacher which shaped several of Dulwich's most famous students, their later work, and our common culture. His students included Chandler, P. G. Wodehouse, and C. S. Forester, all of whom wrote masterpieces celebrating the common man.

We have sought and found our heroes in many cinematic works. Costume dramas are now rarer than they once were (because of the production costs), but the Zorros and Robin Hoods continue to return to the screen. The action/adventure hero is now a mainstay, whether fictional (as in something like *The Guns of Navarone*, 1961) or historical (such as in *The Dam Busters*, 1954). Biography has also been fertile ground for the production company in search of material; we will discuss a number of cinematic biographies in the chapters which follow.

A particularly interesting development—one related to the notion of the common man as hero—is the plethora of successful American films dealing with unlikely heroes. We have also shown our taste for such heroes in the manner in which we have gratefully welcomed foreign examples of this variation.

One thinks of such early examples as *King Kong* (1933), a film with a giant ape as protagonist. We have seen a touching heroine in the form of a literate spider who writes words in her webs in order to save the life of a porcine friend (*Charlotte's Web*, 1973). *The Exorcist* (1973), celebrates a modern priest at war with an ancient demon in Washington, D.C., a city and society not known for its belief in the supernatural. *All the President's Men* (1976) makes heroes of investigative reporters and *Norma Rae* (1979) focuses upon the heroic efforts of a small factory labor organizer.

Some of the recent heroes and heroines have been particularly unlikely: a sprinter who will eventually be a missionary and who will not run on the sabbath despite his opportunity to do so in the Olympics (*Chariots of Fire*, 1981); a Japanese tax collector (*A Taxing Woman*, 1987 and *A Taxing Woman Returns*, 1988); a village theatre operator who teaches a young friend how to operate a film projector (*Cinema Paradiso*, 1988); a competitor for the honor of making the best available fast-food noodles (*Tampopo*, 1986); two lesbian fry-cooks (*Fried Green Tomatoes*, 1991—the lesbianism made more explicit in the novel than in the film); and perhaps most unique of all, a psychotic but extremely polite and well-educated cannibalistic psychiatrist who serves as a mentor to an aspiring female FBI academy cadet (*The Silence of the Lambs*, 1991).

It is clear that we want and need heroes and that Hollywood is anxious to provide us with them. It is also clear that many American cinematic heroes are "common" people or at least unlikely material for heroic status. That speaks volumes about our conception of ourselves as a people and the types of representations which we find resonant.

In summary, it is impossible to provide a brief definition of the hero of the sort that one might encounter in a dictionary. However, there are certain things that one can say that could serve as an outline for such a definition.

The hero is an individual who possesses some or all of a series of relevant *characteristics* (see Figure 4) and who meets certain basic human *needs* (see Figure 5). S/he can appear within a series of contexts (ordinary, national, global, historical) and even move across those contexts. Certain types of heroes arise with greater frequency in some historical eras than others because of the nature of those times. The relevance of such heroes to their contemporaries is conditioned by both the historical time line within which the potential heroes and heroines appear and the personal time lines of the individuals who might choose to consider them heroic. Culturally, we seek heroes who meet both continuing needs and needs that are temporally contingent; at the same time, those who provide representations of such heroes in response to cultural need often shape that culture and that culture's tastes and desires, since there is an ongoing interplay between need and representation. Finally, the hero is an individual (real or fictional) who embodies a pattern of behavior, an individual whose actions are part of a larger experiential arc, an arc which contextualizes the behavior and makes it "heroic" in our eyes.

Chapter III

Heroism and Human Development Theory

> We shall draw from the heart of suffering itself the means of inspiration and survival.
>
> —Winston Churchill

What is the developmental model against which the life of the hero might be measured? If many heroes are developmentally exceptional and hence "abnormal," is there a risk that abnormality might lead to bad results? How are such results to be avoided?

As we commented earlier, our sense of the heroic is affected by our own position within the total stream of human life. We define ourselves, in part, in relation to the historical flow in which our own lives are situated. As the noted scholar of religion, James P. Carse writes, in his book, *Finite and Infinite Games:*

> One cannot be human by oneself. There is no selfhood where there is no community. We do not relate to others as the persons we are; we are who we are in relating to others.
>
> (1986, p. 45)

Cinematic, novelistic, and dramatic representations conceive of human lives as portions of narrative within a larger narrative. Our own lives—lives lived in the so-called real world—are also, actually, stories, the crucial difference being that the ends of our lives, the ends of our stories, remain uncertain, while the lives of fictional characters or the biographies of individuals who are now deceased are rounded and complete. The outcome of our own story is not yet known and we have no way of knowing where we are actually positioned on our personal time line. We know that "today is the first day of the rest of our lives" but the number of days beyond today remains a mystery. In seeing the experiences of others and in observing statistics, we make assumptions about where we are currently positioned, but there is always the chance for a pleasant surprise or sudden disappointment.

Many no doubt tend to overlook or repress the degree of uncertainty in their lives and thus fail to "live in the moment." In part this is because uncertainty causes fear. We may also overlook this uncertainty because we are balanced on a temporal knife-edge. Our entire past is known; our entire future is before us, and the present moment becomes the past the instant we begin to think of it. In some ways it is more comforting to think of human life in terms of fictional or biographical narratives. They are more intense to the extent that our experience of them is more concentrated; hence, they may appear to be more interesting. While they might include touches of day-to-day experience, they are also built largely upon conflict, conflict which seizes and holds our attention. They are not actions but rather, as Aristotle said, imitations of actions. Thus they may sometimes appear to be richer than our own experience, which we know, firsthand, in all of its highs and lows, its excitement and its tedium, its rarity and its sameness.

The 1973 television documentary depicting the lives of the Loud family in Santa Barbara was more like our own in its realism, since it included humdrum detail along with the occasional touch of action and excitement. "Reality" programming such as the show *Survivor* now consists of a very different sort of narrative. That narrative is heavily shaped and edited, distilling dozens of hours of experience into a single brief episode, each episode designed to seize and hold the attention of the viewing audience. We *live* like the Louds (and sometimes experience frustration as a result), but we often think of our personal life story in more dramatic terms.

One way of conceptualizing our frustrations, for example, is to say that things are not going the way we thought (or hoped) that they would. In other words, we have written our own stories in our minds (or at least the next few chapters of those stories) and we are heartily displeased when reality does not choose to cooperate with us.

There are also common expectations which give form to the human story, expectations resulting from demographic statistics or expectations arising from developmental theory. We believe, for example, that children should (and hopefully, will) outlive their parents. When this does not happen, tragedy results. Similarly, we believe that parents should (and hopefully, will) love and nurture their children. When this does not happen, pain and suffering of various kinds result.

Experiential expectations are more easily codified than developmental ones. For example, we can consult actuarial tables in order to draw conclusions with regard to such matters as life expectancy. The media offer us constant "factoids" on how large families are likely to be, how often they are likely to relocate, and what sorts of income brackets they are likely to occupy. While the uncertainties of life are not removed, some of the certainty which we traditionally associate with mathematics (and mathematical descriptions) can be factored into our expectations.

We can consult census and other data and be reassured by the numbers which those data yield. We can determine the likelihood of our dying from a snakebite or an airplane crash, a shark attack or a terrorist's bomb. The RealAge program enables us to factor in our personal measurements and behavior (height, weight, blood pressure, smoking practices, etc.) and estimate the *real* age of our bodies and likely life expectancy. Such knowledge can be comforting, while the unknown is frequently discomforting.

Developmental expectations, on the other hand, are far more complex. We know how many Americans are likely to live past the age of 100 and we know how many are likely (and how many are unlikely) to suffer certain diseases. We do not, however, know such basic developmental things as the relative degree to which our lives are shaped by nature or nurture (though we know that both are important). We do not know the manner in which one's sexual orientation is established. We do not know why siblings raised in the same household under more or less the same conditions by the same parents are so different from one another in personality and behavior. We continue to learn more and more about such issues and we do make inroads against our vast ignorance, although in the case of classic debates such as that of nature versus nurture we often feel that we are simply seeing a swing of the pendulum rather than a steady advance.

Nevertheless, hard data (regardless of whether or not they will apply in our own particular cases) are comforting. They provide a framework (even if it is a rough framework) within which we can think and plan. We know, for example, the age at which individuals commonly marry within our particular society. *Why* they do or *whether* they *should* are other questions altogether and the importance of such practices to our developmental processes are a matter of far more intense debate than the simple underlying statistics which describe or delineate common practice.

Returning to our earlier formulation, we noted that heroes embody *patterns* of experience. Their lives are narratives or, at least, the pertinent heroic episodes are narrative segments. If our own lives can also be said to be narratives, narratives that are as yet incomplete, questions arise concerning the impact of the heroic narrative upon our personal narrative. Does the hero, for example, inspire us by suggesting that our experience (our narrative) can in some way parallel his or hers? Do we ourselves seek to attain some level of heroism by attempting to imitate the patterns which we discern in the narratives of those we admire?

Equally important, can we say that there is some relationship between the heroic narrative and the common human narrative (to the extent that that common narrative can actually be delineated and outlined)? Is there a bit of the heroic in all of us? Is there a confluence between the heroic and the "ordinary" or is the heroic somehow beyond the reach of the average individual?

We have already indicated that we will argue that heroes often diverge from those patterns established by developmental psychologists; they do not, in short, always fit the presumed "template." They may miss or skip developmental stages; they may be seen as "abnormal" and even as deeply flawed. We began with Odysseus' wound—that characteristic by which he would be definitively identified—and would now like to pursue that issue further. Do the hero's wounds irrevocably separate him or her from the rest of us or, on the contrary, does the hero encourage and inspire us because each of us knows our own wounds and flaws and can see, within the memorable context of the experience of the hero, the possibility for our own redemption and triumph?

In pursuing these questions we should continue to remind ourselves of the extent to which fictional or "enhanced" narratives—films of the lives of real people, for example—select and emphasize certain aspects of experience. Such narratives appear more real or more lively than the workaday narratives in which we commonly find ourselves. At the same time, we must remember that heroes often viewed their own lives as we view ours: filled with tedium and frustration. *Most* of us, Thoreau wrote, live lives of quiet desperation. We must, in short, look past the superficial differences between our narratives and the hero's. For example, films could be made of our own lives that would make them appear far more interesting and engaging than they might seem to us to be when they are lived on a day-to-day basis. At the same time, films need not always focus upon the dramatic moments of life. Such films as *Mediterraneo*, *Il Postino*, and *Cinema Paradiso* have been successful because they illustrate the beauty and drama of what might appear to be common, everyday events in the lives of common people with uncommon spirits.

Sometimes a film can reinforce the sense of urgency in our lives by contrasting the protagonist's dreams and desires with the quotidian realities of his life, as, for example, in the film *Patton*, in those scenes between the North African and Italian campaigns and Patton's assumption of command of the Third Army in Europe. His restiveness and impatience underline the sense of frustration we all feel when we believe that banal realities are blocking the way to the fulfillment of our destinies.

If heroes seem somehow "different" from us we should remind ourselves of the fact that we often feel separated (or even alienated) from our own ethos. Each of us endures the trials of the teenager, desperate to feel comfortable within our lives, our communities and, indeed, within our own bodies. We somehow wish to be different from everyone else and "special" even as we wish to be the same as everyone else and "normal."

Harry Berger (1973) notes that our love of certain types of fiction serves to explain our common human needs. We wish to be both exhilarated *and* comforted. We like shocks and surprises, conflicts, adventures, and extremes; at the same time we like rounded conclusions and a clear degree of predictability. We are, in short, both bored and frightened. We seek challenge but we also seek relief. We seek excitement but we also seek peace. We seek a world that is more interesting than ours but also a world that is more fair and more just.

We are prepared to be stretched and jostled but we also wish to see a shape to our experience which can serve to persuade us that, fundamentally, our lives and the larger arc of our experience have some order and some meaning. The template against which our experience can be measured has been the principal subject of the developmentalist's studies. This is, obviously, a vast and complex subject, but we can approach it by using the work of a major developmental psychologist and psychoanalyst as an ongoing illustrative example.

The work of the late Erik H. Erikson has been widely read and widely celebrated. Born in Frankfurt in 1902 and dying in 1994 Erikson virtually spanned the twentieth century, his personal time line paralleling that entire period. Coming to psychology from art, as he notes (1963, p. 18), the aesthetic underpinnings of his psychological thought are often apparent. In the course of his life Erikson was the recipient of both the Pulitzer Prize and the National Book Award. Associated with such institutions as Yale, Harvard, and the University of California, Berkeley, Erikson numbered among his colleagues such individuals as Ruth Benedict, Gregory Bateson, Margaret Mead, Martin Loeb, Alfred Kroeber, and Anna Freud. Although his work can hardly be equated with the total accomplishment of developmental psychology, it clearly represents the most well-known and most widely-acknowledged example of work which focuses upon developmental issues.

While one can here only hint at the breadth of his accomplishment, Erikson's major works include *Childhood and Society* (1950, 1963, 1985), *Identity and the Life Cycle* (1959, 1980), *Insight and Responsibility* (1964), *Identity: Youth and Crisis* (1968), *Dimensions of a New Identity* (1974), *Life*

History and the Historical Moment (1975), *Toys and Reasons* (1977), *The Life Cycle Completed* (1982) and important studies of both Luther (1958) and Gandhi (1969).

Erikson's stage theory has been widely discussed and praised. While alternatives and adjustments have been suggested by such scholars as Lawrence Kohlberg (1963), who stressed the individual's *moral* (as opposed to psychosocial) development, the notion of a stage theory—the belief that our lives progress through a set of predictable episodes—is widely accepted and the acceptance of that broad notion is more germane to our purposes than specific details or minor points of conflict.

Erikson's position can be briefly summarized. While accepting the basic tenets of Freudian theory, Erikson feels that Freud's emphasis upon the psychosexual should be complemented by an emphasis upon the psychosocial. (Where Freud, e.g., would concentrate on the sexual dimensions of something like orality, Erikson would focus on its social dimensions, addressing such matters as the place of communication within personal relationships.) Erikson speaks of eight psychosocial stages which encompass the human life span and discusses the fact that each individual's life cycle exists within a larger community of others and their life cycles, the interactions between the members of the community being essential for both individual development and the shaping of human culture. This emphasis upon the social plays to the nurture side of the much-debated nature/nurture relationship, whose debate was of special importance in the early days of developmental psychology, particularly in the work of Arnold Gesell (1948, 1956, 1972).

Erikson is particularly known for his studies on the development of identity and for his efforts to develop therapeutic methods which look beyond the more structured psychoanalytic setting generally used with adults (Miller, 1989, pp. 176–177). Erikson's developmental model can be seen as something of a compromise. It is neither as tightly integrated as Piaget's nor as loosely connected as Freud's (Miller, 1989, p. 180). While Erikson's focus falls upon the quest for *identity* as a unifying theme in developmental theory he does not overlook issues which receive special stress in the work of other developmentalists, for example Kohlberg's emphasis on the practice of moral reasoning as a key aspect of the individual's growth and maturation (Lerner, 1986, pp. 267–273).

In general, it can be argued that Erikson's approach includes central elements of the thought of each major developmental theorist. His particular prominence—attributable to his longevity, his prolific writing for a wide and diverse audience, and his renowned biographical studies—make him a particularly appropriate model against which to measure the conclusions of this study.

As we have noted, stages are also of considerable importance to Freud. Indeed, the mindset which seizes upon *evolutionary* process is characteristic of the nineteenth century, both in Darwinian science and in the period's overarching, optimistic ideology of an upward trajectory of human accomplishment. The idea of "progress," so often associated with the Victorians, actually takes hold in the renaissance (Jones, 1961) in debates concerning the age of the earth. While one side holds that we have regressed since former, "golden" ages and are now in decadence and desuetude, the other argues that we are actually in our youth, with all the world before us. If there were once giants upon the earth (including intellectual giants) we now build upon their work. We stand—the common aphorism states—upon their shoulders.

Evolution is not without violence, as Darwin argued, but while the debates over evolution have themselves sometimes been violent, evolution's existence (whether in Darwin's precise formulation or not) is no longer in doubt. Along with debates with regard to the ways in which evolution can still accommodate the notion of divine creation, some of the more interesting debates in this regard concern the extent to which human progress can keep in step with (or fall behind) technological progress. We can now, for example, perform medical procedures (involving genetic engineering, for example) that may well be in advance of our human capacity to absorb those procedures' results and implications. Similarly, while we have proven ourselves capable of improving technology, we have been far less successful in improving human beings. Torture, one of the major concerns of enlightenment thinkers (Beccaria [1764], for example) and a vivid example of human barbarism, continues to be practiced despite indictments of it that are now hundreds of years old. Indeed, it has become the subject of a succession of recent, popular (i.e. lucrative) films, such as *Saw* (2004) and its many sequels.

Nevertheless, we *have* changed. Child labor laws have been enacted in many industrialized countries and women do now vote in most. Women are less frequently considered chattel than they once were. It is interesting to note that women often refer to more "progressive" men as being more *evolved*. Clearly then, change can occur. It can and does occur across human society and it certainly occurs within our individual lives, the larger patterns of cultural change being mirrored in the microcosmic development of the individual. Gail Sheehy's immensely successful book *Passages* (1976) demonstrates the popular acceptance of the essential need for human change and development and the pivotal notion that that development occurs across a series of *stages*.

To return to Erikson, the following schematic outline (Figure 6) represents a summary of his stage theory. Erikson's well-known model includes eight

Figure 6. Erik Erikson's "Worksheet" Summarizing the Eight Stages of Development.*

	Psychosocial Crises	Radius of Significant Relations	Related Elements of Social Order	Psychosocial Modalities	Psychosexual Stages
1	Trust vs. Mistrust	Maternal persons	Cosmic Order	To get; To give in return	Oral; Respiratory; Sensory-kinesthetic
2	Autonomy vs. Shame and Doubt	Parental persons	"Law and Order"	To hold (on); To let (go)	Anal-urethral; Muscular (Retentive-Eliminative)
3	Initiative vs. Guilt	Basic family	Ideal prototypes	To make (=going after); To "make like" (=playing)	Infantile-Genital; Locomotor
4	Industry vs. Inferiority	Neighborhood, School	Technological elements	To make things (competing); To make things together	Latency
5	Identity and Repudiation vs. Role Confusion	Peer Groups and Outgroups; Models of Leadership	Ideological perspectives	To be oneself (or not to be); To share being oneself	Puberty
6	Intimacy and Solidarity vs. Isolation	Partners in friendship, sex, competition, cooperation	Patterns of cooperation and competition	To lose and find oneself in another	Genitality
7	Generativity vs. Self-absorption (Stagnation)	Divided labor and shared household	Currents of Education and Tradition	To make be; To take care of	None
8	Integrity vs. Despair	"Mankind"; "My Kind"	Wisdom	To be, through having been; To face not being	None

*From Erik Erikson, *Identity and the Life Cycle,* 1980. Adapted from Patricia Miller, *Theories of Developmental Psychology, 1989.*

stages within the developmental processes of the individual. The child begins in a world of conflict between trust and mistrust. He or she is dependent upon a primary caregiver. Generally this would be the child's mother. Beyond the mother is a panoply of uncertainties and threats. The adult closes his or her life (hopefully) with an intact ego and a feeling of wholeness and satisfaction rather than a feeling of despair.

The steps between these stages are not absolute ones. Some individuals, for example, might experience all of the stages but never fully move beyond one or more of them. For example, an individual might live his or her entire life without fully moving beyond basic mistrusts. Similarly, an individual might move beyond a particular stage but then fall back after the occurrence of some traumatic incident. If, for example, an individual is significantly betrayed in middle age he or she might fall back into a mode of distrust, a mode from which he or she might never recover.

In Erikson's model each developmental stage is accompanied by a particular virtue and a particular strength (see Figure 7). For example, in the generativity stage, an individual exhibits the strength of productivity and exemplifies the virtue of care. One of the hallmarks of Samuel Johnson's achievement as a writer and as a man was the concern which he expressed for young writers, writers whose careers he helped advance long after his own was established. For Erikson this expression of "generativity" is a sign of maturity, an indication that someone is progressing across the developmental stages. When we are older and our careers are defined and settled we should not fall into stagnation. On the contrary, we should help the next generation to come into its own. We should serve as mentors and helpmates. Moreover, we should find joy in that process. Mature individuals within that stage of human development should want to help those who come after them. In doing so they also exhibit strength and practice virtue.

The function of *virtue* here is an important one, one to which we will return. Here, however, we would note the fact that not everyone progresses through these processes in the same way. Johnson could be cited as an appropriate example. The subject of the most famous biography in English, Johnson is also the most frequently quoted writer in English after Shakespeare. Known for his wit and wisdom, he is a hero to many.

Johnson's life, on the other hand, was anything but orderly and tidy. He was born half-dead, he said (Johnson, 1958); he was such a pitiful sight that most individuals would not have picked him up off the street. His mother and father quarreled frequently, principally over money, and he overheard them. While aware of his many talents, his mother forced him to perform before strangers, a practice he avoided by hiding from her whenever he thought she might wish to put him on exhibit.

Figure 7. Erik Erikson's Stage Theory Model.

DEVELOPMENT PROCESS OF THE INDIVIDUAL	BASIC VIRTUES	STRENGTHS
1. Trust vs. Mistrust [Approx. Birth to Age 1]	Hope	Drive
2. Autonomy vs. Shame and Doubt [Approx. Ages 2 to 3]	Willpower	Self-Control
3. Initiative vs. Guilt [Approx. Ages 4 to 5]	Purpose	Direction
4. Industry vs. Inferiority [Approx. Ages 6 to Puberty]	Competence	Method
5. Identity and Repudiation vs. Role Confusion [Adolescence]	Fidelity	Devotion
6. Intimacy and Solidarity vs. Isolation [Young Adulthood]	Love	Affiliation
7. Generativity vs. Self-absorption (Stagnation) [Middle Adulthood]	Care	Production
8. Integrity vs. Despair [Late Adulthood]	Wisdom	Renunciation

Adapted from: Erik H. Erikson

Note:

Joan Erikson adds a Ninth Stage which she terms *Gerotranscendence*, in which the challenge is to sustain the successes achieved by negotiating the other eight stages in the face of diminished physical and mental capacity. In rising above these challenges by "doing and making do" our last years become a *gerotranscendance*.

Joan M. Erikson, *The Life Cycle Completed, Erik H. Erikson*, 1998.

Johnson had a younger brother, about whom we know very, very little. He died young and it is possible that he was prosecuted for a white-collar crime, perhaps forgery. Johnson says next to nothing about him. Moreover, despite his praise for his mother Johnson avoided seeing her for years at a time, even when the visits would have posed no great problems of time or expense.

When Johnson married he married a widow who was nineteen years his senior. She had social pretensions and a fondness for opium; we know that for a time they were separated. We also know that while she was alive Johnson kissed and fondled a younger woman who lived within their household. Johnson and his wife Elizabeth had no children.

We know that one of Johnson's earliest memories was his mother's account of heaven and hell, a fire-and-brimstone story which he never forgot. Religion was always a source of fear for him as well as a source of comfort. While he was an infant, suffering from scrofula, he was taken to London to be touched by the Queen. Scrofula, or tuberculosis of the lymph glands, was referred to at the time as the "king's evil." It was believed that a monarch could heal the afflicted individual with a single touch because of his or her possession of what was then believed to be "divine right." Thus, Johnson was taken to the capital to be touched by Queen Anne.

Regrettably, Anne's touch did not cure him, but it is clear that the experience had a deep impact upon him, because Johnson remembered it all of his life. At one point during the journey he and his mother were travelling by coach and Johnson became sick, upsetting the other passengers. He also remembered that his mother feared being robbed and kept some of her money sewn in her petticoat. He was given some silverplate to remember the experience and it was engraved with the name Sam J. rather than the simple initials S. J., because his mother's name was Sarah and in the event of her death unscrupulous individuals might try to take the silver from him, claiming that it was the property of his mother. Later, when they were suffering financial distress, he and his wife were forced to part with it.

The point of these anecdotes is that while Johnson was a great man and remains a hero to many, his life did not proceed along a neat, simple, and straight line. His memories were replete with the effects of superstitions and false fears. His abilities were, early on, a source of pain and embarrassment for him. There were great rents within the fabric of his family and problems in his marriage. Throughout his life he was plagued by melancholy. As a youth he would go on long walks between Birmingham and his native Lichfield to try to conquer his depression. After the completion of his dictionary he suffered what has been described as a nervous breakdown (Bate, 1975). He emerged from that state with the help of the Thrales, a wealthy London brewer and his wife who provided him a place within their family. What he needed, W. J. Bate argues, was the stability of a middle class family. Fortunately the Thrales were there to provide it.

Thus, Johnson's life was a difficult one, wracked with psychological turmoil, poor health, and an array of emotional setbacks and conflicts. He suffered from shame and guilt. Measuring himself within the context of his own capabilities rather than the capabilities of others, he often considered himself to be lazy and unproductive. He suffered isolation and stagnation. He did not, in short, progress smoothly across Erikson's stages.

At the same time, he was able to accomplish many things and achieve the status of hero in the eyes of many. In addition to writing poems, essays, sermons, and myriad other works, he completed four major projects: the cataloguing of the massive library of the Earl of Oxford, the editing of the first English dictionary (it had taken the forty members of the French Academy forty years to compile theirs), the editing of the complete works of Shakespeare, and the editing of the works of the English poets from the renaissance to his own time, with biographical prefaces to the works of each poet. He lived seventy-five years, surviving scrofula, deafness in one ear, blindness in one eye, stroke, enlarged heart, emphysema, and a host of other maladies. On his deathbed he refused opiates, saying that he wanted to meet his God with a clear head.

Herein is the center of this study—the extent to which one can achieve heroism without successfully negotiating standard developmental stages. One might go so far as to suggest that certain individuals might even have achieved their heroism in part by their failure to successfully negotiate those stages. When one door closed, another opened. Frustrated emotionally, Johnson put his enormous energies to work on great projects. His feelings of loneliness and isolation drove him to the company of men and women such as Reynolds and Hawkins, Hester Thrale and Boswell, who left records of his great conversations, sharing them with society and thus preserving his legacy and insuring a portion of his immortality. Failing as a schoolteacher in the midlands he left his wife to set off for London, where he literally gave his name to an age.

What then is the relationship between heroism and human development, or, more precisely, what conclusions might we draw from the experience of heroes *vis à vis* the assumptions of developmental stage theory? And more specifically, how can the breaches of *normality* in many heroes' lives lead them to great achievement rather than to disaster?

We can approach the issue usefully by coming at it from an opposing angle. In 1992 Robert K. Ressler published a book entitled *Whoever Fights Monsters*. The book was long awaited, since Robert Ressler was the official at the FBI who was responsible for the psychological profiling of special criminals, principally serial killers. These are *the* individuals whose lives, like those of many heroes, involve significant breaches of normality.

It was Ressler whom Thomas Harris consulted as part of his research for his blockbuster bestseller, *The Silence of the Lambs* (1988). Harris' book is, of course, only one of many to deal with serial killers and the FBI's profiling system, one of the later of the better known books being Patricia Cornwell's *The Body Farm* (1994). Serial killers haunt the contemporary imagination and

Ressler was responsible for providing significant insight into the reasons why serial killers do the things that they do.

In the course of his work Ressler studied the activities of such individuals as Richard Trenton Chase, William Heirens, Charles Manson, Richard Speck, Ted Bundy, Jerome Brudos, John Joubert, Gerard Schaefer, John W. Hinckley, Duane Samples, Harvey Murray Glatman, John Wayne Gacy, Edmund Kemper, Monte Rissell, and Jeffrey Dahmer. Obviously, each individual is different, but in studying the cases of these particular perpetrators of heinous crimes Ressler found certain repeating similarities.

One of the most common experiences of this type of individual was childhood neglect. Commonly left to themselves (or left to themselves with concomitant experience with other forms of physical, sexual, or psychological abuse) these individuals were unable to develop normal affective relationships. In fact, they did not understand normal affective relationships since they had never experienced them.

Moreover, as a result of their neglect, they had often hardened themselves psychologically as a defense or survival mechanism. Thus, they became incapable of normal feelings, of normal emotional responses. Since they were incapable of relating to members of the opposite sex in normal, socially-approved ways, they often developed alternative approaches based on their personal fantasy lives. Having been abandoned and isolated they inhabited worlds of their own making rather than interacting within the world of loving (or, at least, engaged) parents and siblings.

Since they had inured themselves to the pain of abuse and abandonment by repressing their feelings, they were unable to experience normal emotions and feelings within affective relationships. Thus, they pursued extreme, heinous behaviors for three principal reasons: they did not understand normal relationships; their own views of human interactions had come from a world of fantasy; and their inability to feel in normal ways required that they pursue extreme actions so that they could move beyond their repressed emotions and reach a level where they could be able to actually *feel* something. Lacking normal empathy, for example, they would have to inflict grievous pain on others before they could reach the point at which they themselves began to feel alive and engaged. (Stuart L. Brown [1994] has identified parallel patterns of deviance and violence which he has associated with certain childrens' inability to develop normal patterns of play.)

If one were to characterize Ressler's serial killers' experience within a developmental context, one would say that they had not sufficiently negotiated the stages of human development through which "normal" individuals—to one degree or another—pass. In fact, they were incapable of negotiating the stages because of the pathological nature of their emotional lives.

Thus, like the lives of some heroes, their lives did not follow a common human pattern (or at least follow a common human direction). However, while an individual such as Samuel Johnson could transcend his developmental disabilities—his inability to negotiate certain developmental stages—and become a hero to many, these individuals were unable to negotiate developmental stages and instead became serial killers.

Thus, it is clear that while the inability to move effectively from stage to stage across the developmental spectrum may have some bearing on the fact that certain individuals can move beyond those experiences and achieve great things, other individuals with developmental disabilities are unable to do that and instead end up as, literally, monsters. Why? We would argue that the defining difference between the hero and the monster—each of whom may fail to follow the standard developmental pattern—lies in the acquisition of *virtue*.

Returning to Erickson's chart (see Figure 7) each stage in the developmental process has a concomitant strength and virtue. What we intend to suggest in the chapters that follow is that while one may "fail" the developmental stages, one may *not* "fail" the virtues. If one possesses the virtues associated with the various stages one can survive the developmental disabilities that may interrupt and mark one's life. Indeed, one can do more than simply survive. One can become heroic. It is not enough to simply have emotional scars and developmental disabilities; that, after all, is also the lot of the serial killer.

One might offer another example here to illustrate the point, an example which has been as resonant with popular readers and viewers as Harris' depictions of serial killers. In 1976, after a long career as a Professor of English at the University of Chicago, Norman Maclean published a creative-nonfictional account of his youth and young adulthood entitled *A River Runs Through It*. The book recounts a number of episodes in his family life in early twentieth-century Montana.

Essentially, the book deals with two brothers—Norman, of course, and his younger brother Paul. The two grow up in a family dominated by their father, a Scot and Presbyterian minister with a special fondness for fly fishing. Their father teaches his sons the gospel as well as the techniques of the fisherman and along the way notes a special relationship between the two.

The family is a very close one. Norman, for example, is educated at home by his father. The book ends with Paul's death. He dies as a result of violence and the cause of his death (along with the identity of the perpetrator[s]) remains unknown. It is likely, however, that Paul has failed to pay off a gambling debt, for the murder is unusually violent—his casting hand is crushed-

-perhaps indicating that the perpetrator(s) wished to use Paul's death as a message to others.

The entire book is a preparation for the conclusion. Paul is the family favorite. His father describes him as "beautiful," a word which echoes across the book in a host of ways. Norman and his father speak of Paul's death briefly and then never mention it again. The one word which is remembered from their conversation is that single adjective. Paul and Norman's mother is not even there. Overcome with grief she has withdrawn to her room. Norman and his father sit quietly, trying to understand their brother/son, and being unable to do so, characterize him in a way that for each is both appropriate and deeply felt.

The striking thing about this "ordinary" hero is that no one truly knows him. He remains an enigma to both his friends and family and fails to establish a continuing, committed relationship with anyone else. Unlike his brother Norman he does not marry and raise a family. He never leaves Montana. He works as a newspaperman, but he remains a thrall to his youthful activities of drinking and gambling. Put simply, he does not grow up. He stays frozen in time, a part of the landscape, like the rivers and trout streams that haunt Norman's memory.

There are two things for which he is never late: working and fishing. He may stay up the entire night before, but he is faithful in these two areas of experience and utterly dependable. As a fly fisherman he is an artist. He develops the method of so-called 'shadow casting', which serves as the inspiration for both the cover illustration of the mass market paperback version of the book and the marquee posters advertising the film.

Paul and Norman never really communicate in the manner of most adults. Instead, they fish together, positioned apart at various points of the river. They are "together" in a very limited sense of the word. Physical proximity and the sharing of a particular activity lead to a form of intimacy which is neither verbal nor, finally, communicable to others.

As Norman's audience we learn very few things: Paul is young and, by standard measures, immature. He does not grow up. For all of his brief life he pursues the activities of youth. In one area, fly fishing, he is superb, and he is so superb that the adjective chosen to describe him is *beautiful*. He is, in short, a hero, at least to his father, mother, and brother, a hero who has touched all of their lives, a hero of sufficient proportion and dimension to serve as the center of a book and a film which continue to reach a wide audience.

Paul is developmentally disabled; he is also, in some measure, a hero; he is also the possessor—with his weaknesses—of some identifiable virtues. Within that area of his life that really constitutes his heroic life (his fly fishing), he has hope, willpower, purpose, competence, fidelity, love, care, and wisdom, the full gamut of virtues specified by Erikson.

What Norman Maclean demonstrates in *A River Runs Through It* is that we need not pass through each developmental stage to achieve something memorable. Moreover, we can achieve something memorable at the same time that we are beset by clear failings and weaknesses. We achieve something memorable when we are in possession of virtue. We need not be understood by others; we need not even be capable of being known by others; what we must do, however, is embody a pattern of behavior sufficient to be judged heroic, a pattern which is marked by the practice of identifiable virtues.

Maclean's brother's notoriety is now associated primarily with his elder brother's account. We would never have known of Paul Maclean if it were not for his brother Norman's book, a book that touched the director, Robert Redford, and depicted a character of sufficient interest and importance to attract Brad Pitt. This private hero has become a public one and has touched all of those who have met him (though—like his family—not really known him) through his brother's art.

Chapter IV

Heroes Real and Imagined

> There are new words now that excuse everybody. Give me the good old days of heroes and villains. The people you can bravo or hiss. There was a truth to them that all the slick credulity of today cannot touch.
>
> —Bette Davis

Theory is all well and good, but where are the examples?

The test of any theory comes with the adducing of supporting data. Are there really wounded individuals who might be characterized as developmentally disabled who nonetheless rise above their pain (in some cases using that pain as leverage) and become heroic models for the rest of us? There are indeed. There are so many, in fact, that we will restrict our extended examples to a single theme and then offer a longer list of individuals whose experience also fits this pattern.

The theme that we will use is *connectedness* and its absence. It is sometimes said that the greatest source of pain in the modern world comes from *alienation*—alienation from one another amidst urban sprawl, alienation from nature as we walk on concrete and asphalt rather than on the grass and earth, alienation from ancient faiths as religion is subjected (often successfully) to skeptical attack, alienation from society as we are polarized by single-issue voting, alienation from our government as the private lives of our leaders are exposed by investigative journalism, alienation from our families as the incidence of illegitimacy continues to rise and the rate of divorce remains high, alienation based on gender, alienation based on the displacement of human space by cyberspace, and alienation from ourselves as new neuroses and syndromes are precipitated by a world of white noise and constant change.

We desperately want to belong, to feel connected, to feel anchored, to feel at home. While change has always been a human constant, as Heraclitus long ago noted, change now comes more quickly and more insistently. This provides opportunity, but it also poses significant challenges. How have individuals coped with the feeling that they are somehow disconnected? How have they achieved stasis and success in the face of circumstance? How have they taken their own experience of disconnectedness and created success and achievement?

A contemporary example.

In the fall of 1998 Joyce Maynard published a book entitled *At Home in the World*. Described as a memoir by its author, reviewers quickly turned it into a controversial book, some even considering it a *succès de scandale*. The core event of the book is Maynard's revelation of the details of her relationship (at age 19) with the reclusive author, J. D. Salinger, then 53. Maynard was accused, by some, of invading Salinger's carefully-guarded privacy by revealing those details. Alternatively, one might argue that a middle-aged writer anxious to develop a sexual relationship with a person who is also a writer (Maynard had published a piece in the *New York Times* that elicited Salinger's 'fan' letter that inaugurated their relationship) should expect (unless he is hopelessly naïve) to see some of the details of that relationship later appear in print.

In point of fact the Salinger episode is only one segment of Maynard's memoir, which actually consists of an account of a series of relationships—principally with her parents—that might have resulted in truly debilitating alienation. Her father, for example, was an alcoholic. A very gifted man, he never really made a mark in his profession (he was an English professor at the University of New Hampshire) but instead dabbled in painting. Her mother, an equally gifted individual, had an odd relationship with her daughter that involved bedtime activities which she described as 'snuggling'. Later in her life, when Joyce married and had children, the marriage ended in divorce. She is now a single mother who supports her family through a successful writing career.

Her memoir *At Home in the World* is an excellent book, one of the most interesting and engaging of recent autobiographical reflections. It does exact a certain amount of vengeance on Salinger, but that seems fully deserved in light of the circumstances recounted and no one has come forward to counter the facts as she presents them. What has not been commonly noted is that her debunking of the Salinger mystique has revealed him as one of the great phonies of our time, a real term of derision for Holden Caulfield, so that

Maynard has, in effect, rewritten *Catcher in the Rye* and turned Salinger into one of the antagonists.

The book is much more than that, however, and the point here is that a young woman who has experienced dysfunction and abuse in multiple forms has taken that experience and built a successful career upon it. In doing so she has given heart to many and achieved a sense of personal equilibrium which she describes as feeling "at home in the world." In short, she has transcended multiple forms of alienation.

Her experience recalls that of another famous contemporary, James Ellroy, whose mother was the victim of an act of random violence in a homicide case that has never been solved. This event (and others within his family) led him to a dysfunctional life of crime and substance abuse, from which he eventually emerged. Now one of the top two or three crime writers in America, Ellroy has recounted the experiences connected with his mother's death (*My Dark Places*) and written a tetralogy dealing with crime in Los Angeles (the site of his mother's death—actually El Monte). He has, in effect, built a writing career on his personal, painful experience. In doing so he has achieved both material success and personal peace.

Mark Fuhrman, the LAPD detective accused of perjury and planting evidence in the O. J. Simpson trial, was forced to leave his position in Los Angeles and move his family to northern Idaho, far from the relentless pressure of the media and the charges that were leveled against him. The distinguished writer and prosecutor, Vincent Bugliosi, has studied the Simpson case and the charges against Fuhrman in depth. He concludes that it is inconceivable that Fuhrman planted evidence and that Fuhrman's racist epithets (which cannot be excused) were always directed against criminals and that his various partners (several of them African-American) never detected any racism directed at black people as a group. Moreover, as recently as October 1994, Fuhrman had passionately dedicated himself (successfully) to exonerating Arrick Harris, an African American accused of murder, a fact that the prosecutors failed to introduce in Furhman's defense at the Simpson trial (Bugliosi, pp. 172–173).

Bugliosi also assesses the degree of 'perjury' in Fuhrman's testimony at the Simpson trial (p. 419); in his judgment Fuhrman's denial that he had used a racial slur within the previous ten years (which he had used) does not rise to the level of actual perjury, because this was not, in Bugliosi's judgment, material to the guilt or innocence of Simpson. Telling an untruth under oath does not, in and of itself, constitute perjury. The untruth must be material to the case.

Bugliosi concludes that Fuhrman was a true victim in the Simpson case (and he does so, it should be noted, in a book replete with the outrage mentioned in the title, outrage directed at the prosecutors, the defense attorneys, the judge and the media). Nevertheless, Fuhrman was at fault for uttering the slurs and he suffered dearly for it. Lacking the money to defend himself against the resulting perjury charge, Fuhrman abandoned his career with the LAPD, accepted a conviction, gathered up his family, and left the city.

While he is ineligible to hold a private investigator's license his skills in detection are still strong. After seeing his life nearly destroyed by a case in which those skills were neither fully utilized nor appreciated, he took those skills—which had been challenged—and utilized them as a writer.

To date he has produced three significant true crime books, one on the Simpson case, one on the Martha Moxley case, and one on the case of a serial killer in Spokane. His book on the Moxley murder was so compelling that the case was reopened after decades of inactivity. Moreover, the individual identified as the likely murderer by Fuhrman, Michael Skakel, was eventually convicted of the crime. Skakel, the nephew of Ethel Skakel Kennedy, did not—like Simpson—escape justice through the power of money and celebrity.

Fuhrman's bestselling books highlight his skills as an investigator and have enabled him to demonstrate his tenacity and determination to a national audience that only knew him through the distortions presented by the Simpson defense team. His life in ruins and his ability to support his family in serious jeopardy, Fuhrman turned the source of his wounds into the source of his eventual triumph, attaining not just material success in the form of bestselling books, but justice in the form of a conviction of an individual who had, like Simpson, previously escaped such justice.

Rather than accepting the fate of a person who constituted little more than a forgotten footnote to what Bugliosi and many others consider a significant miscarriage of justice, he struck out, afresh, in search of justice and eventually attained it. In telling the story of others he has, in a sense, told his own story as well, for he has demonstrated the skills, the objectivity, and the integrity which had previously been challenged. He has done so with relatively little fanfare and with exceptional professionalism.

There is a far more prominent writer, however, whose personal experiences were often limited and fragmentary, but whose effect on the world has been tremendous, a writer whose world and culture are very distant from our own but a writer whose reputation has never been higher: Jane Austen.

Describing and understanding a world of which one is barely a part.

Jane Austen's life is far simpler on the surface than her art. Born at Steventon Rectory in Hampshire in 1775, she died in Winchester at the age of

41 in 1817. The daughter of an Oxford-educated, rural clergyman, Jane never married. Nor was she able to support herself by her writing. After the death of her father, the Reverend George Austen, she—along with the other women in her family—were dependent upon her brothers for their support. Her brother Frank brought her, her mother, and her sister Cassandra to Southampton after her father's death and burial at Bath in 1805 and in 1809 her family eventually moved to Chawton Cottage, a modest but adequate house in rural Hampshire provided by her brother Edward from his inheritance of the estate of Thomas Knight.

Edward had been permitted by the Austens to be adopted by the childless Knights, a practice not uncommon at that time. The wealth which he received from his adoptive parents enabled him to provide subsequent help for his mother and sisters. This sometimes comes as a surprise to modern readers, familiar with Jane's later fame and success. However, while she enjoyed some income from her writing, it did little more than enable her to purchase a few London luxuries such as chinaware. Her writing did not provide her a livelihood or anything approaching financial independence. Without the help of her brothers the family would have been in very difficult circumstances. As it was, their circumstances were quite modest.

Chawton Cottage is a small, two-story brick structure in a rural village. Everything about it is modest in size, including the reception areas, the dining room, and bedrooms. Jane is said to have done most of her writing in the dining room. The writer some consider the greatest novelist in history had no study or formal library. When the family moved from Steventon to Bath, earlier, they had to sell the bulk of their possessions because they could not afford the expense of moving them a relatively modest distance.

It is interesting to note that while she was one of the greatest writers of fiction in the English language her tombstone in the North Aisle of Winchester Cathedral makes no reference whatsoever to her writing. She is remembered as a loving sister and faithful, virtuous Christian, a true and apt description, but an incomplete one.

It comes as a surprise to many modern readers that she often concealed the fact of her writing, the squeaky steps and flooring of Chawton Cottage alerting her to the arrival of others so that she might quickly put away her writing materials. While writing was a central part of her life, it was not the only part of her life. She had time for many other activities, including the encouraging of her niece in her own efforts as a writer. Though she had a sharp wit and a discerning eye she was particularly remembered for her fundamental decency and lovable nature. By outward signs her life was quite uneventful. While her love of a young man named Tom Lefroy was to some degree reciprocated (he traveled to England upon news of her death and also purchased a literary memento associated with her [a letter from the publisher Thomas Cadell]), their financial condition made marriage impossible. In December of 1802 she received a proposal of marriage from a gentleman named Harris Bigg-Wither. She accepted it, but then changed her mind the following morning after a sleepless night. There are unspecified accounts of a love match with another man, but his sudden death ended the relationship and the letters concerning the man were destroyed by Jane's sister, Cassandra (Nokes, 242–44). The presence of love in any marriage was central to her attitudes and beliefs. "Any thing is to be preferred or endured rather than marrying without Affection," she writes. "Nothing can be compared to the misery of being bound without love, bound to one and preferring another."

Thus, her decision to decline a proposal and remain unmarried was a studied and conscious one. She remained single and lived in relatively narrow circumstances by choice, passing from the dependence on her father to the later reliance upon her brother. One has some sense of the obscurity in which she lived by visiting Steventon, a tiny village accessible by the narrowest of roads that is little more than a set of lanes. Driving from Steventon to such assembly sites as The Vyne, a local, stately home, one has a sense of how narrowly circumscribed was the world in which Jane Austen dwelt.

The original versions of both *Pride and Prejudice* and *Sense and Sensibility* were written by the age of 22, while she still lived in Steventon. This indicates how considerable were her gifts. Moreover, the issues raised in both novels reflect her own experience with regard to the system of primogeniture and the resulting legal/financial plight of English women. The manner in which British inheritance law discriminated against wives and daughters is central to the plot of each novel. Thus, while she and her family suffered as a result of that system, she addressed it, critically, in her fiction, without allowing it to destroy her good humor and kind demeanor. She did not, in short, permit herself to be obsessed with any perceived victimhood. At the same time, she did not turn a blind eye to the shortcomings and problems inherent in the system which was responsible for her own experience and the experience of a host of others. Basically, she took one aspect of her plight and utilized it as a subject for successful narrative and trenchant commentary. She combines a belief in the necessity for social stability with a realization that that stability also requires searching criticism, an attitude which she shares with another writer who commanded her great respect: Samuel Johnson, the writer whose monument in Westminster Abbey is right above her own.

In the end, she sought love rather than comfort; she chose spinsterhood over loveless security, and happy obscurity over unhappy prominence. While these choices were not without some regrets and not without some occasional bitterness they were choices by which she was prepared to stand. In her work we see her own displaced dreams, depicted with great poignancy, in part because she herself was not able to achieve those dreams. Anne Elliot, the heroine of *Persuasion*, who is extremely close to Jane in personality and aspiration, eventually marries Captain Wentworth, becoming a sailor's wife. Jane can write of this action with warmth and authority, for two of her brothers became admirals in the British Navy. She can also write with concern and sympathy, for she herself having limited options for travel spent all of her life in a few counties of southern England. She writes of a love and of a body of experience which she herself sought but never enjoyed. The compelling narrative is born in dreams, in pain, and in despondency—particularly during her time in Bath, as Claire Tomalin (1997) has noted. Bath, a spa town, is a site for holidays and assemblies, for high spirits and high living. For Jane it was a place of sadness. *Persuasion* is set in Bath and the plot turns on the possibility of dejection and even anguish turning to happiness.

It is imprecise to say that Jane Austen suffered from developmental 'disabilities.' Jane Austen chose what she considered to be higher goods over what others might have considered "normal" goods. She also left a novelistic legacy in which she explored questions of personal and social justice and the manner in which those questions affect the lives of sympathetic protagonists, their sisters and mothers.

In her novels there is often a struggle between virtue and inclination (Elizabeth Bennet is prejudiced, Darcy prideful) and a demonstration of the fact that virtue can succeed in the face of straightened economic circumstances and/or the deleterious impact on the individual of the practices of common law and common custom. Virtue is, indeed, frequently rewarded and its opposite frequently punished. Individuals are endowed with free will though they function within the constraints of society and as they utilize their free will virtuously they may be rewarded in spite of the power exercised and pressure exerted by that society. Elinor Dashwood weds Edward Ferrars in *Sense and Sensibility* and Captain Wentworth marries Anne Elliot in *Persuasion* after enduring considerable pain and negotiating numerous moral hurdles. Emma Woodhouse marries Mr. Knightley only after she has been chastened by him for her thoughtlessness and arrogance.

There is also an ongoing appreciation of the fact that those who are "different" from us may also be superior to us as well as inferior to us. She neither patronizes nor romanticizes the poor and she neither worships nor envies the rich. Wealthy aristocrats and once-wealthy aristocrats are criticized in both *Pride and Prejudice* and *Persuasion*, just as they are presented with love and sympathy in *Emma* and *Sense and Sensibility*. Insensitive relations are criticized in both *Pride and Prejudice* and *Sense and Sensibility* while loving relations are presented very positively in both books. The poor are treated sympathetically in *Persuasion* and negatively in *Sense and Sensibility*. All comes down to individual morality, individual behavior, individual choice.

Jane Austen respects independence (as well as wit and candor), even if her society was often less comfortable with them than she was and she shows how they can lead to success as well as to temporary impasses. What is particularly striking about the current interest in her work is the fact that her novels project a world in which the consequences of actions and choices are kept at center stage. It is a world with moral expectations as well as cultural expectations, a world in which moral actions are rewarded and immoral actions are condemned rather than "understood," a world in which the moral compass is always operative and always pivotal for the assessment of character (and for the prediction of the individual character's ultimate reward or punishment).

In short, Jane Austen's is a world with little tolerance for moral relativism. It is a world with codes and expectations and its resonance today suggests that we have been hungering (at least to some degree) for the advantages of such a world. We have also been hungering for the clarity of such a world (without oversimplifying the complexity of contemporary experience). There is a *moral* "normality" in Jane Austen's world which is not yoked with a demand for *behavioral* normality. Lizzy Bennet can walk when others ride, she speaks when others would remain silent, and she refuses to dance when she believes she has been criticized unfairly. Moreover, she can pass up a fortune of £10,000 a year when the marriage proposal she receives is laced with criticism.

She is, in a very real sense, rewarded for her independence, but that independence is virtuous. She is a good and decent person, honest, loyal, and courageous. What she requires is a similar demonstration of virtue from others and the best of them are attracted to her (Darcy in particular) because the degree of her independence and the extent of her virtue mark her character. She does not fawn. She is no sycophant. She is polite, but very much her own person. She reads when others play cards. She roughhouses with Bingley's dog when others are pursuing more conventional activities. Her remoteness from the norm is what seizes Darcy's attention. It captivates him. It also arouses his passion.

We mentioned above Jane's high regard for Samuel Johnson. Johnson's situation is quite different from Jane's, though his conclusions are similar and his intellectual orientation comparable.

As noted, above, Johnson's early years were difficult ones, particularly with regard to the relationships within his family. Thus, when Johnson writes (in *Rasselas*, in his periodical essays, in his verse and other prose) of the difficulties faced by families and households, he does so with authority grounded in a lifetime of experience. It has been said that his insights and observations constitute the greatest anticipation of the thought of Freud before the twentieth century (Bate, 1955, p. 93, and see Grange, 1962). From his often painful, day-to-day experience, comes a lifetime of wisdom and a focusing upon the real lives of real people in real situations which looks forward to the later flowering of both biography and social history, whose importance he argued for throughout his life.

Johnson's experience was broader than Jane Austen's, but narrow by modern standards. He traveled in Wales, in Scotland, and in France, for example, though the writer who Johnson would have singled out as the great student of human nature and human life—Shakespeare—spent all of his life in Warwickshire and London. Writers such as these reach down into themselves for the experience and insight which they transmute into great art, art capable, ultimately, of changing a world they could not foresee and extending to corners of the earth which they never saw.

Their writing—built upon their experience—served to make them whole and complete as individuals, even in the face of the pain which often constituted that experience. In their healing is our inspiration, inspiration emerging from imaginative works of art and inspiration emerging from our perception of their ability to transcend both the limitations and pain of their experience and instead, structure a body of work upon it, work which has endured for centuries and—particularly in the case of Shakespeare and Jane Austen—has never been more compelling for a broad, popular, indeed global audience.

Finding oneself amid the crowd.

Thomas Edward Lawrence (1888–1935), Ned to his family, Lawrence of Arabia to the post-Lowell Thomas world, T. E. Shaw or T. E. Ross to some of his friends and military companions, presents a particularly interesting case for the student of heroes and heroism. Lawrence is, by turns, a very real man as well as a very important media creation. John Mack writes:

> A poet friend wrote to Lawrence once: 'You are a kind of double personality for me (a) a charming friendly character given to sudden appearances on inordinate motor-bicycles; (b) a sort of mythological figure or force of nature—so that I feel as if I were being addressed by an earthquake, in however still and small a voice.'
>
> (1976, p. 459)

Lawrence's celebrity results from his demonstrably real strengths as well as from a series of quite fortuitous circumstances, circumstances which concern cultural perspective as well as military necessity. We spoke earlier of the importance of luck for the hero; in this regard Lawrence is a textbook case.

While the principal focus of our interest concerns the developmental lessons that one might draw from Lawrence's life, the cultural lessons are no less interesting.

The David Lean film, *Lawrence of Arabia* (1962), introduces Lawrence to us as an adult. After an opening sequence concerning Lawrence's apparently accidental death in a motorcycle accident, we cut to Arabia and Lawrence's adventures there. As with Patton and Churchill, it is interesting to note the manner in which filmmakers edit out certain portions of their lives, even when those portions might be of uncommon interest.

In the case of Lawrence, a later sequence provides a brief account of his earlier life. We learn, for example, that Lawrence's family name is really Chapman, not Lawrence, and that his mother and father did not marry. Actually, Lawrence's mother was his father's housekeeper/governess, Sarah Junner. While the "illicit" love of his Victorian parents yielded eight sons, with five surviving into adulthood (Montague Robert, T. E., the second eldest, William George, Frank Helier, and Arnold Walter), it also brought a great deal of guilt, particularly for his mother.

In addition to other relocations in the face of negative public opinion, Lawrence's father moved the family to Oxford in order to secure the best education for his sons. What we do not learn in the film is one of the key elements in Lawrence's development, namely, the fact that Lawrence proposed to a woman (Janet Laurie Hallsmith) who declined his proposal, accepting instead that of his younger brother, Will. However, Will was killed in World War I (as was Lawrence's brother Frank) and Lawrence was left with a lost brother, a lost love, and the permanent competition of an undying memory. Lawrence shared with this woman a sizeable portion of his inheritance and they remained close friends, though she chose another man to be her husband. Thus, his earliest and longest memories of human intimacy were suffused with guilt (in the case of his parents) and loss (in the case of his would-be fiancée). Perceiving the pains that attended attempts at intimacy and being frustrated in his own attempts to attain it, he would later find himself and establish an important part of his identity in his relationships with groups. As he sought contacts with groups he avoided contacts with individuals. Thus, the developmental gaps in his experience helped make him a memorable figure on the world stage.

The fact that his family had been uprooted, combined with his personal inclinations, yielded a childhood marked by an active curiosity and a rich fantasy life. Shortly after the family moved to Oxford, in 1896, Lawrence explored a shed at the bottom of the family garden. In it he found a pipe which intrigued him. In order to explore it further he took a hammer and nail and pierced it. Water gushed forth and his father was forced to call in a plumber. His brother Bob commented (Yardley, 1985, p. 24) that this curiosity was so developed in him that by his teens he was the one to whom everyone turned for the diagnosis of mechanical problems and the mending of broken objects.

At the Oxford high school he was a natural learner, "one of few words," a master said, "self-possessed, purposeful, inscrutable" (Yardley, 1985, p. 24). He was bored by the daily routine at school and avoided organized games. In physical appearance he was like his mother—small, but with a muscular body. Instead of competing on the playing field he proved his worth to himself on his own terms, following harsh, self-inflicted rules. He developed a reputation for an iron will and, in the words of his school friend, E. F. Hall, a "ruthlessly fast-driven brain" (Yardley, 1985, p. 25).

He especially enjoyed cycling, an activity encouraged by his father, one of the sport's earlier enthusiasts. Cycling enabled him to escape the rigid confines of home and school and follow his youthful interests, wherever they might lead. He was particularly fascinated by the past and he would cycle for miles around Britain, visiting castles, churches, and ancient buildings. Moreover, his father had taught him to take photographs even before his fourth birthday. In the summer of 1906 he and a friend set off on a cycling tour of the coasts of Normandy and Brittany. Lawrence stayed for more than a month, making extensive notes on each castle and historic building he visited. In the spring of 1907, having passed his scholarship exam for Jesus College, he spent two weeks in Wales, again on his bicycle, returning to France later that summer.

He also read widely and especially enjoyed epic romances, particularly those dealing with the Crusades. He personally identified with Richard I, one of England's most romanticized monarchs. He indulged in brass-rubbing at every opportunity, the subjects of the rubbings often being medieval knights. His Oxford tutor, Reginald Lane Poole, remembered that Lawrence's essays were always controversial and frequently relied on romantic passages from medieval writers (Yardley, 1985, p. 32). The manner in which he followed his own lights gave him the reputation of an eccentric. He neither smoked nor drank, for example, and rarely ate meat.

That he should associate himself with chivalric figures was entirely appropriate. The Chapmans were actually related by blood to the family of Sir Walter Ralegh, the famous Elizabethan courtier, writer, and adventurer. In the introduction to his edition of Lawrence's letters, David Garnett notes (1938, p. 37) that when Lawrence was nine years old he used to tell his brothers a story night after night without end. Long sections were in rhyme. The tale concerned the successful defense of a castle against various violent foes. The heroes of the story were the brothers themselves.

(Lawrence's experience finds its parallel in the life and work of C. S. Lewis. Lewis' mother feared that the Belfast of their time might infect her sons with typhus, typhoid, or cholera and required Jack [C. S.] and his brother Warren to remain in their house for uncommonly long periods of time. Unable to play outside like other boys, they indulged a rich fantasy life in the family attic, where they drew pictures and told endless stories about the imaginary country of 'Boxen'. What Lewis lost as a child helped prepare him for a literary life which included the Narnia stories. Lewis' books now sell nearly two million copies a year.)

At Oxford Lawrence said he spent most of his time reading the imaginative literature of medieval France, particularly the long heroic poems called the *gestes*. He noted that Galahad was the illegitimate son of Elaine and Lancelot—the purest of knights and the one for whom the grail would be reserved. As an illegitimate son himself, the story was especially important to him. He also liked a thirteenth-century French epic called the *Huon de Bordeaux*, in which one of the bravest of the knights is named Gerard, an illegitimate son of Charlemagne. It has been noted that Eric Kennington's stone effigy of Lawrence in Wareham Church, Dorset (originally intended for St. Paul's and, later, Salisbury Cathedral) quite consciously imitates the carved effigies and brasses of medieval warriors. In reproducing a picture of the effigy for his edition of *Crusader Castles* (Lawrence's undergraduate thesis) Michael Haag terms it "The Last Crusader." Chapter I of Liddell Hart's book on Lawrence, *The Man Behind the Legend*, is entitled "The Crusader."

Having won his scholarship to Jesus College, Oxford, Lawrence was finally able to utilize his abilities and merge them with his natural inclinations. Fortuitous experience helped him early. Just prior to 1906 some medieval pottery had turned up in excavations for a pub in the Cornmarket in Oxford. Lawrence's interest was immediately aroused. From that moment on he occupied himself with examining the deep foundations which were unearthed and he paid the workmen to keep any little pieces which they might find. These he later donated to the Ashmolean Museum, where he established a relationship with its Junior Assistant Keeper, Leonard Woolley and its Keeper, David Hogarth. This passion for antiquities was constant. On one occasion he and his friend, C. F. C. Beeson, were caught in the crypt of St. Cross Church, examining human bones.

The research for Lawrence's undergraduate honors thesis on crusader castles required 2,000 miles of travel across Europe. He also worked with Hogarth and later Woolley, excavating the mound of Carchemish, an ancient Hittite city. In addition to his work at the Ashmolean, Hogarth was in British Intelligence and Lawrence became his recruit.

His knowledge of Arabia clearly enabled him to become the hero in fact and in legend that we know today. Carchemish was the direct link between Lawrence's student days at Oxford and his wartime adventures in the Middle East. The opportunity to exhibit military prowess and physical courage would never have existed had it not been for his academic prowess, his direct, personal knowledge of the area, his facility with language, and his deep sense of history. He was, in effect, an "expert," a specialist whose knowledge could be tapped and exploited by his military superiors and by the politicians with whom they were forced to conjure.

However, there is far more to the Lawrence of Arabia story than the confluence of personal abilities with public events. Lawrence is a public figure, in good measure, because of the efforts of the contemporary press. The story is a fascinating one and, again, involves fortuitous events.

In his book *The Adventurer* (1974), Paul Zweig draws a distinction between the *hero* and the *adventurer*. The hero fights with a clear moral purpose; any battles in which he is involved are battles which pit good against evil. Such a man is courageous, loyal, and, above all, selfless. The adventurer, on the other hand, acts out of a desire for the pleasure and exhilaration of the adventure itself. He may, of course, be a hero as well (like Lawrence), and he is certainly not (by definition) without moral purpose. However, he *seeks* adventure, while the hero may find it, just as the adventurer might *find* himself in a war with evil, where the hero consciously agrees to risk his own life and happiness in that transcendent battle.

Zweig argues that the public appetite for stories of adventure (both fictional and non-fictional) was greatest in the 1920's and 1930's and he adduces as key examples of the quintessential adventurer T. E. Lawrence, Charles Lindbergh, and the pulp novel hero, Doc Savage. It is interesting to note that both in the David Lean film and in his own writings, Lawrence refers to his efforts in Arabia as *fun*. His actions there grew out of his personal intellectual interests but those interests were combined with a real and important taste for adventure. Lawrence realized early on that had he not taken part in the Arab revolt he would have sought some other activity that would take him beyond the bounds of the university. While prepared to devote his time and energies to the pursuit of scholarship, he was not prepared (like his brothers Will and Arnold) to circumscribe his life by accepting an academic appointment.

At the same time, of course, his public life was marked by a unique dimension: that of the solitary man within a larger group. After the war Lawrence would join (under pseudonymous names) the Royal Air Force and the British Tank Corps. He wanted to be part of something larger than himself and he was willing to alter his identity in order to do so. In effect, he was replicating his father's own experience, sacrificing his name and, in a sense, his identity, for higher purposes.

In Arabia, however, the solitary man was a leader of a group, a fact that was brought to public attention by a rather circuitous route. Lawrence of Arabia was, in part, the creation of the journalist, Lowell Thomas. On the instruction of the U. S. government (and with the financial backing of a group of Chicago businessmen), Thomas was sent abroad with the mission of stirring American enthusiasm for the First World War.

Witnessing the brutality and desolation of the trench warfare on the western front (James, 1990, pp. 275–276), Thomas was taken in hand by John Buchan (then Director of the new Department of Information) and guided to the Middle East and to Lawrence. Buchan, a renowned novelist, known particularly for his book *The Thirty-Nine Steps* (1915), was a friend of Lawrence; one of the characters in his Hannay novels, "Sandy Clanroyden," has in fact a number of features in common with Lawrence. Thus, a real person who had already been portrayed fictionally is recommended to an individual in need of a real-life example to portray propagandistically.

In Lawrence, Thomas found exactly what he sought: a modern *crusader* for the liberation of the Holy Land and the emancipation of its human communities. In the process he also discovered a great deal of human interest, depicting "Britain's modern Coeur de Lion," as Thomas termed him for an eager audience. Lawrence was the perfect hero: freed of the necessity to deal with the anonymous carnage of trench warfare on the western front, Thomas had an *individual* hero—a man set against the open desert landscape rather than plunged into the muck of foxholes, trenches, artillery craters, and barbed wire of Europe; a *bona fide*, recognizably romantic hero, a man whose exploits would resonate for an American audience that had already welcomed the Valentino movie versions of the pulp desert sagas of E. M. Hull. (One might also mention a later, cognate cultural example, Sigmund Romberg's popular operetta *The Desert Song* [1926], which featured a romantic, desert hero saving the good from the forces of evil and again illustrating the public taste for romantic heroes silhouetted against a desert sky, preserving all that is good and true.)

Thomas' books and public performances lionizing Lawrence were extremely effective. In addition to the "American" dimensions of his heroism—the western-style individual set against the expanse of an endless blue sky—Lawrence was admired for his abilities to utilize gadgetry. Lawrence, it was said, was a master with explosives and he would not share the secrets of their use with his Arab comrades. The tulip bombs which he used to disrupt the Turkish railway system were less sophisticated than the later weaponry of

James Bond, but technical skill in the use of ingenious devices has long been a feature of the modern hero in his most popular manifestations.

Again, luck merges with circumstance. It is unlikely that (Colonel) Lawrence would ever have found himself at the head of the Arab revolt were it not for his interest in languages and the history and archaeology of the Middle East. However, with the help of Thomas, Lawrence's experience is transmuted into semi-myth and reinforced by his cultural heir (Commander) James Bond, himself a student of languages (though Bond is a Cambridge man, rather than an Oxford man like Lawrence).

This cultural continuity is not a matter of accident. In fact, the cultural adaptation of certain heroic norms for twentieth-century adventurers is a matter of considerable interest. It points up the interplay between fiction and fact and indicates how certain traits, behaviors, and activities became associated with heroism.

The link between Lawrence and James Bond is the aforementioned Doc Savage, the creation of a mid-west telegrapher by the name of Lester Dent. Dent's early pulp writing caught the attention of Dell publishing, which brought Dent to New York and positioned him to write (as far as can be determined) at least 165 Doc Savage novels.

Doc Savage was termed the "Man of Bronze" because of the deep tan he acquired in the course of his adventures in warm, exotic locales. Like Lawrence, Doc is frequently seen in broad, exterior settings. Whether at sea or in the air, Doc served as an optimistic alternative to American readers caught in the throes of the Depression, longing for a distinctly American hero possessing both a taste for adventure and the opportunity to indulge it.

Doc Savage was also a master of gadgetry (Dent had created another pulp adventure-hero with the rather more straightforward name of "Gadget Man.") He also possessed a so-called fortress of solitude in the frozen north, to which he would repair for the study of scientific articles. In effect, he went there on sabbatical in order to stay current with his field. Here, in short, is an inheritor of the Lawrence mantle who also served as the prototype for two of the most important twentieth-century fictional heroes—Superman (the Man of Steel [not bronze]) and James Bond. Doc Savage's surname was Clark and Lester Dent's pen name for the Doc Savage series was Kenneth Robeson. Clark . . . Ken/Dent . . . Kent.

Doc Savage offered many of the same features as Lawrence: courage, skill, a sense of adventure and fun, a certain discomfort in the presence of women, a refusal to be hemmed in, an interest in science and scholarship, and the ability to exhibit his abilities on a broad stage against a set of exotic backdrops. He also possessed an American sense of fair play and basic decency. A kinder, gentler hero, Doc employed "mercy bullets" which temporarily rendered his adversaries helpless so that they could be deprogrammed/brainwashed and returned as wholesome contributors to society. In the depression, it was felt that many putative "criminals" had been forced into a life of crime and with a little help from goodhearted friends they could be transformed back into normal, decent Americans. That was part of Doc Savage's mission.

Interestingly, Lester Dent's own experience was not unlike Lawrence's (and, indeed, Savage's). Marilyn Cannaday's biography of Dent (1990), to which we are indebted for much of this information, takes note of the fact that Dent himself was an anguished loner. A native of La Plata, Missouri, Dent's family moved to the wilds of Wyoming where they were separated by many miles from their nearest neighbors. Until his time in school (which required his mother to move away from their farm and into a distant town to accompany him) Dent was virtually without human contact, save that of his immediate family.

Dent had no sense of how to relate to women. His marriage resulted from his placing an ad in a newspaper which expressed his willingness to teach anyone who was willing to learn how to operate a telegraph machine and send messages in Morse code. Dent was simply lonely and in need of human companionship; his future wife answered the ad and they struck up a friendship. Instead of living in society, Dent had spent his life up until then living within his own consciousness, reading of knights and battles, studying science and pursuing scientific hobbies.

Later, of course, this all bore fruit in Dent's pulp writing and the money he earned (as much as $18,000 one year in the depths of the depression) enabled him to travel and to purchase three airplanes and a forty-foot schooner. These experiences and possessions contributed to his ability to write adventure stories but they also physically separated him from "normal" society. Eventually he returned to La Plata and built a home which he designed personally. At a time when many individuals did not even bother to lock their homes, Dent installed a radio-activated security system. He also developed a primitive but effective garage-door opener and residents of La Plata would sit on the street outside of Dent's house hoping for an opportunity to see him park his car. To them, he was strange, bizarre, and withdrawn, but quite fascinating.

Many personal parallels could be drawn between Dent's isolation, his lack of human interaction, his sense of adventure, and striking accomplishments, and Lawrence's. They are of a different order and sometimes of a different nature, but the parallels remain interesting. The correlation between abnormal development and later achievement, for example, is clear. In the introductory chapter to *Seven Pillars of Wisdom* (1935), Lawrence writes that:

> All men dream: but not equally. Those who dream by night in the dusty recesses of their minds wake in the day to find that it was vanity: but the dreamers of the day are dangerous men, for they may act their dream with open eyes, to make it possible. This I did.
>
> (p. 23)

Some heroes exist in the pages of history. Some exist in the pages of fiction. Some (like Lawrence) exist in both and influence real men who create fictional heroes who then influence other real men and other creators of fiction. The interplay between fact and fiction is ongoing, as is the cultural influence of these heroes and superheroes.

A 1927 "boys' book" by Bridges and Tiltman, entitled *Heroes of Modern Adventure*, claims that "every boy is an adventurer at heart," and then goes on to feature T. E. Lawrence as the subject of one of its chapters. The ultimate adventurer, the model adventurer-hero in the west, is Odysseus and it is not surprising that Lawrence would choose as his last major work (though he was not aware that it would actually be his last) a prose translation of the *Odyssey*. There Lawrence scores Homer for his lack of experience. "Few men," he writes, "can be sailors, soldiers and naturalists. He was driven by his age to legend, where he found men living untrammeled under the God-possessed skies" (1932, pp. 8–9).

In *The Mint*, Lawrence's account of a certain period of his life in the R.A.F., he refers to his enlistment as a "voyage into the unknown." This was a different voyage, however, than his voyage to Arabia and he concludes (p. 231) with a reflection on "how man hangs in his body, crucified." Like Yeats, whose poetry Lawrence admired, man is seen as an adventurous spirit locked within the body, fastened, as Yeats says in "Sailing to Byzantium," "to a dying animal" (l.22). Daydreamers age and their lives change. They continue to live in consciousness though their bodies cease to cooperate with them in their search for the answers to their hopes and aspirations.

In the dedicatory verse to *Seven Pillars* Lawrence wrote, "I drew these tides of men into my hands/and wrote my will across the sky in stars . . ." but the Nietzschean will of youth ultimately gave way to the bitter realities of barracks life in *The Mint*, where Lawrence did not enjoy the opportunity to lead a great number of men in a desperate battle but, rather, encountered the day-to-day routine of men—many of them crude in word and action—in the peacetime army.

Lawrence's commonplace book, *Minorities*, is a collection of poems, presumably his favorites, drawn from the 300 volumes of verse in his library at Clouds Hill, Dorset. He has de la Mare on Arabia there, Coleridge's "Kubla Khan," the great romantic poem of adventure which contrasts with the far darker "Rime of the Ancient Mariner," Morris' "Sir Galahad: A Christmas Mystery," and Sir Walter Ralegh's poem "His Pilgrimage." Perhaps the most interesting inclusions are lines from Shelley's "Hellas":

> The world's great age begins anew,
> The golden years return . . . .
> A loftier Argo cleaves the main,
> Fraught with a later prize;
> Another Orpheus sings again,
> And loves, and weeps, and dies.
> A new Ulysses leaves once more
> Calypso for his native shore.
>
> (1824, ll. 1060–61, 1072–77)

Lawrence is, in so many ways, a great adventurer, particularly the Lawrence of *Crusader Castles*, *The Wilderness of Zin*, the translation of the *Odyssey*, *The Mint*, *Minorities*, and key passages in the *Letters* and *Seven Pillars*. Writing to his mother, for example, in 1908, Lawrence again quotes Shelley:

> I love all waste
> And solitary places: where we taste
> The pleasure of believing all we see
> Is boundless, as we wish our souls to be.
>
> (Garnett, ed., 1938, p. 56)

The *adventurer* Lawrence need not be opposed to the *hero* Lawrence, though it is the adventurer side which often seizes the attention of contemporaries and linked Lawrence with similar figures within the period of the 1920's and 1930's. It is a side which emerges naturally from his intellectual experiences and interests and it is far more romantic and attractive to think of war as adventure than as bloody work. ("It's not just a job, it's an adventure," the television ad claims.) Moreover, in certain ways, Lawrence is time-bound. If Lawrence James (1990) is correct with regard to Lowell Thomas' intentions and Lawrence's later actions, Thomas took real materials but created a semi-mythic figure from them, a figure whose outline Lawrence himself helped to shape and construct in the course of his interactions with Thomas.

Instead of a complex but reasonably straightforward example of developmental disability leading to greatness (a man incapable of personal intimacy who can only find love and identity within a group), what we now have is a complex and conscious interplay between private life, public history, and popular consciousness. The "identity" was not only found; it was also systematically shaped. And were it not for historical circumstance it might not have been found at all. While he was prepared to take on his role, Lawrence needed something as grand as a world war, as all-too-human as the hype of a Lowell Thomas, and the precise confluence of time and opportunity to truly become "himself," realizing that that self is part reality, part fictional construct, aided and abetted for the purpose of attaining real political and personal goals.

For many, the quintessential image of Lawrence comes in the second half of the David Lean film. It is the image which is used in many of the promotional pictures concerning that film. Lawrence and his Arab followers have just destroyed the tracks of a railroad with a tulip bomb. The train leaves the tracks; Lawrence fires a flare down the line of his troops, stands up, his back against the sky, and calls his men to follow him for the attack. They rise up and do so as he stands before them, white robes flowing in the desert wind. The importance of the physical setting and its contrast with the dark worlds of such World War I films as *Paths of Glory* (1957) or *All Quiet on the Western Front* (1930) speaks volumes. In the Lean film Lawrence is then questioned by a reporter whose role stands in fictional parallel to the real-life role of Lowell Thomas. He, in effect, stands ready to construct the Lawrence image and Lawrence stands ready to help him do so.

Despite the dimension of romance and media manipulation which are part of the Lawrence saga there is also, beneath the surface, a reality that makes the romance and the construction of legends possible. As Liddell Hart writes,

> Legends are more potent than emperors or dictators. Others who worked with him were outstanding men: he would have been the first to wish their merits due recognition. Legend has made his fame as 100 to 1. Such magnification, which happens to a few men in each generation, is not true to reality. But it is true of this case that legend had a solid basis, of far greater content than usual. There will be nothing but good in it, if his real message is remembered, and not merely the romance.
>
> For he was a message to mankind in freedom from possessiveness. In freedom from competitiveness. In freeing oneself from ambition, especially from the lust of power. His power sprang from knowledge and understanding, not from position. His influence is a living growth—because it is a spiritual message transmitting a spiritual force. The man was great: the message is greater.
>
> (1934, p. 390)

We will return to these matters later, when we consider Churchill's reflections on Lawrence. Here, however, one may conclude by pointing to Liddell Hart's central notion. Lawrence inspired. Lawrence's force was spiritual. Lawrence's strength lay in his knowledge and in his virtue. Returning to Erikson's formulation, it can be said that whatever failings Lawrence possessed, he did possess hope, willpower, purpose, competence, fidelity, love, care, and wisdom. The fact that he suffered from shame and guilt, the fact that he was subject to role confusion and had difficulties with human intimacy are a part of his experience. However, they not only did not keep him from achieving heroism; in some ways they contributed to that heroism.

Today, Lawrence stands as one of the most fascinating and revered of twentieth-century figures. Books concerning him—including small books published locally for a committed, nearly cult-like audience—continue to multiply. Like Johnson and, for example, Churchill, his life is written and rewritten year by year. His spirit haunts our imagination. The outline and arc of his experience haunt our cultural memory. His human flaws may be deep but his virtues are always clear. Had there been no great war and had there been no Lowell Thomas, there still could have been a great man, a hero to a smaller but no less inspired group of individuals privileged to have had the opportunity to know him. We can only dream of the role he might have played, had he lived to participate—with Churchill—in the war effort against Hitler.

From virtual abandonment to Parliament Square.

Winston Churchill (1874–1965) was born in Blenheim Palace, in Woodstock, Oxfordshire on November 30, 1874. This was an entirely fortuitous circumstance. Blenheim has been the historic seat of the Churchills since the eighteenth century, but only for the heir to the dukedom. Churchill's father was not the Duke of Marlborough; his wife was simply visiting Blenheim at the time of Winston's birth. The episode is, however, an interesting and important one. The birth room at Blenheim is still preserved and—replete with Churchilliana (letters, photographs, portraits, and items of personal clothing)—it is a major tourist mecca, particularly since Churchill, his wife, and many members of their family are buried in the Bladon churchyard, a brief distance from the palace.

Throughout his life Churchill was haunted by history. He was, of course, himself a historian and the line of history—both national and familial—in which he found himself was such as to make the strongest of impressions. Later in his life he wrote a study of the family of John Churchill, the Duke of Marlborough who was the first occupant of Blenheim, a work that formed the basis for the popular PBS series, *The First Churchills* (1969).

Blenheim is a magnificent palace, designed by Sir John Vanbrugh, the architect of Castle Howard, Yorkshire, another grand structure, associated in our cultural memory with the successful television series, *Brideshead Revisited* (1981). Sometimes referred to as the English Versailles, the library alone at

Blenheim is over 100 feet long and terminates in a magnificent pipe organ. Interior shots of Blenheim were used in the John Goodman film *King Ralph* (1991) to suggest the scope and style of Buckingham Palace (not available to production companies, of course) and the terrace in front of the main building was large enough to contain a complete movie set for the Covent Garden scenes in the recent remake of the film, *Black Beauty* (1994). Kenneth Branagh's *Hamlet* (1996) was filmed there as well.

The palace was a gift to John Churchill from Queen Anne for his successes in the War of the Spanish Succession (1701–1713). This war was precipitated by the occupation of Spanish fortresses in Holland by French troops, Louis XIV wishing to put his grandson on the vacant Spanish throne. This, of course, would have radically upset the balance of power in Europe. Thus, a large coalition army was formed against the French, the British army being commanded by John Churchill, the first Duke of Marlborough. Churchill enjoyed a succession of military triumphs, one of them at the village of Blenheim, for which the Oxfordshire palace was later named.

The palace is filled with history. The magnificent ceilings, e.g., were done by Nicholas Hawksmoor, the principal architectural assistant to Sir Christopher Wren. In addition to some decorative pieces, ironically enough obtained from Versailles, master portraiture, including works by Peter Lely and Joshua Reynolds, the palace also includes a set of grand, silk tapestries, commissioned by John Churchill. These floor-to-ceiling tapestries depict and commemorate his military victories. There are also such military memorabilia as actual French standards which Churchill received as part of the process of surrender.

The park in which the palace is set—Woodstock Park—and particularly the lake on its northern side were the result of the work of Lancelot "Capability" Brown, the most famous landscape architect in English history. A generation before John Churchill, during the Restoration, the Earl of Rochester, John Wilmot, the second most important poet of the period after Dryden, was Ranger of Woodstock Park. Thus, it is impossible to be in that particular setting and not see, feel, and actually immerse oneself in English history, particularly that period of English history when the English people were both triumphant in art and triumphant in war.

The first *Winston* Churchill, a seventeenth-century ancestor of the first Duke of Marlborough, is depicted in a small portrait at the palace. This Churchill is associated with the family motto: *Faithful but Unfortunate*. Winston Churchill was a supporter of the executed Stuart king during the civil wars of the seventeenth century and though his side eventually triumphed with the restoration of Charles II in 1660 the parliamentarians won the wars and, as a result, Churchill lost his property.

This is not unimportant, for though the Dukes of Marlborough have been in possession of a great house for over 250 years they have not always been in possession of the funds to maintain it. In (our) Winston's time, for example, one of the dukes (Charles, the ninth duke) was forced to trade his title for an infusion of American capital. He was wed, by arrangement, with Consuelo Vanderbilt, a beautiful young woman whose family was enamored of English titles. After presenting him with two sons—referred to as the heir and the spare (years before the phrase was associated with Princess Diana)—she obtained a divorce and lived happily ever after with a Frenchman. The pull of Blenheim, however, is strong and Consuelo is buried near Winston in Bladon churchyard.

While Winston was to become the leader of the Conservative Party, John Churchill, the first Duke, was associated with the Whigs. Hated and vilified by such Tories as Jonathan Swift, it is ironic that his descendant Winston would be the standard bearer of the descendants of the Tories. It is quite appropriate, however, that both John and Winston Churchill would share another experience in common: both were leaders in momentous events involving attempts by British enemies to conquer Europe and make it their own.

John Churchill, unlike the seventeenth-century Winston and some of his descendants in the dukedom, did not lack money. The fact that Blenheim, a great but private home—is open to the public for a fee of £17.50 and the fact that it is regularly used for such purposes as the production of commercial films, indicates that the resources to support the estate must now be found through such enterprises. The beautiful park, which contains a butterfly house and garden maze is connected to the main building by a miniature railroad. Blenheim is not quite a theme park but neither is it a seat of pure, isolated wealth. It is also an educational site, with a wide array of programs for school groups of all ages.

Thus, it is no surprise that our Winston, for all of his titles and accomplishments, like many of the Churchills, was hard pressed to secure the financial wherewithal to support his style of life. Basically he worked as a prodigiously productive writer, an enterprise and activity which contributed to both his erudition and his eloquence.

He was also, of course, something of a hybrid, an Englishman to the core, but the son of an American mother. He was later granted honorary American citizenship by President Kennedy, but his ties with America had always been strong and his resourceful, entrepreneurial nature had long been in evidence. Blenheim, however, was always in his blood. The lake which sits beneath the great monument to the first duke contains an island planted with a line of towering poplars. It is often said that the poplars are so positioned as to suggest the masts of a great ship, a symbol of British maritime power. Twice in his career, Churchill would be named First Lord of the Admiralty.

Eight miles away from Blenheim is Oxford. Driving to the city along the Woodstock Road one passes Polstead Road. Near the corner, on the north side of the street, stands the Oxford house of T. E. Lawrence, who served under Winston. Thus, the ties and the associations are many and for an individual such as Winston they are of great importance. Winston grew up surrounded by a historical flow of which he himself was eventually to become a part. Between the Palace of Westminster and Westminster Abbey, in Parliament Square, stands a great statue, commemorating him and his accomplishments. Similarly, though he is buried in Oxfordshire, there is a prominent monument in the Abbey.

Churchill has been called the greatest man of the twentieth century; he is certainly one of the most important. In leading his own country during the Second World War he also provided courage and inspiration to the citizens of other countries. Anyone who studies his conduct of the war, anyone who visits the cabinet war rooms (the so-called command "bunker" in King Charles Street) and studies the procedures in place there, realizes quickly the degree to which his involvement in the war effort was direct, personal, and constant.

There was, of course, a Churchill before and after the war. In 1895, working as a journalist, he covered the guerrilla warfare in Cuba and a few years later participated in the Boer War, where he was captured—events covered in the film *Young Winston* (1972). In 1900 he was both recommended for the Victoria Cross and elected to Parliament. In 1905 he became Colonial Under Secretary and three years later he was promoted to cabinet membership. In 1910 he was named Home Secretary, in 1911 First Lord of the Admiralty. In 1913 he founded the Royal Naval Flying Corps and after the outbreak of the First World War he commanded the defense of Antwerp. In 1916 he led a battalion with the rank of lieutenant colonel. In 1919 Churchill was named Secretary for War and Air, in 1921 Colonial Secretary (where one of his advisers was T. E. Lawrence).

In the early 1920's after several electoral defeats he became Chancellor of the Exchequer. In the early 1930's he was instrumental in building the intelligence network in London and Berlin. Throughout the 1930's he warned his countrymen of the threat posed by the Nazis, opposing the official policy of appeasement. In 1939 he was again named First Lord of the Admiralty. After Hitler's invasions of Poland (1939), Denmark, Norway, and France (1940), Winston Churchill was named Prime Minister.

Modern students are sometimes surprised to learn that he was voted out of office at the end of the war, though he was returned in 1951. As dazzling as his public career surely was (regardless of one's political persuasion), it is paralleled by an impressive series of more private accomplishments. As we indicated earlier, Churchill was a prolific writer, producing voluminous accounts of the war, of the Churchill family, and a widely-read, multivolumed *History of the English Speaking Peoples*. He was also a loving husband and father and a skilled painter.

By education he was a soldier. Hence he was educated, not at Oxford or Cambridge, but at Sandhurst. His career was that of a politician and political leader, but his livelihood often came from his writing and his knowledge and erudition were far beyond what one would normally encounter in the barracks or in the hallways of power.

He was, in short, a very great man and the signs of his greatness are never far to seek. In his time and after he was a great subject for mimics and parodists, particularly in his sonorous tones and in such personal mannerisms as the wearing of top hats and the enjoyment of outsized cigars.

One must achieve a significant degree of recognition before one becomes grist for the mimic's art. In modern London Churchill's face is everywhere—in pubs, on souvenir items, in documentary films. He adorns Toby mugs and character jugs and his personal associations with the palace are part of the tourist draw at Blenheim. William Manchester entitled his multi-volume biography of Churchill *The Last Lion* (1983, 1988). Given the symbolism of the lion, Manchester has associated him, inextricably, with England itself, as, indeed, so many others also have done. He is a crucial part of our collective memory of twentieth-century Britain. We have mentioned his monuments in Parliament Square and in Westminster Abbey but his general prominence is clear and his image is everywhere.

There is another Churchill besides the public figure and the writer: a sometimes lost and insecure youth and an elderly man beset by depression. Churchill's parents, Lord Randolph and his American bride, Jennie Jerome, were anything but solicitous. It is not too strong to say that, for a considerable period of his youth, Winston was essentially abandoned. His letters to his parents, desperately soliciting their love and attention, are filled with sadness and pathos. This is so pivotal a part of his early experience that such letters are even on public display at Blenheim.

During this time the most important person in his life was his nanny, Mrs. Elizabeth Anne Everest. Tempting psychological interpretations forever after, Churchill referred to her always as "Woom" (sometimes interpreted as an abbreviated version of *woman*). Until he was twelve years old she was the dearest person in his life (Manchester, 1983, p. 116) and her picture hung in his bedroom until the day he died.

The neglect of his parents—notable even by Victorian and Edwardian standards—had lifelong effects. Such a lack of love and attention not only makes the individual desperate for such attention; it also makes him wonder whether or not he is actually worthy of love. If his own parents do not appear to love him, perhaps he is simply unlovable.

For some, the result of such an experience is distress and, potentially, psychological disaster. For others, including Winston, there is an attempt to win affection and attention through actions. These may not all be positive. In his youth, for example, he was sometimes contemptuous of authority, a fact that brought problems in school. He saw himself as "troublesome" (Manchester, p. 118) and tended *not* to blame his parents. As Manchester argues, a child such as Winston is desperate for attention and affection because he receives so little. The less he receives the more he cherishes what he does receive and he hopes for more. He does not blame his parents, since he cannot bear to confront the possibility of their loss. Thus, he blames himself and idealizes them, seeing them as he wishes they were rather than as they are.

Winston's early troubles in school must come as a considerable surprise to those who associate him with knowledge, wit, erudition, and a rhetorical skill which enabled him to vie with the great speakers and orators of the past. At Harrow, e.g., Churchill got the lowest marks in the school and he was only admitted to Sandhurst on his third attempt.

He also suffered from physical shortcomings. He was slight in stature and was continually beaten and bullied at school. He spoke with a lisp and he also stuttered. At Sandhurst, as Anthony Storr notes (1988, p. 9), he was only five feet, six and a half inches and his chest measured a bare thirty-one inches. By Sandhurst standards this was quite inadequate.

At his second preparatory school (Storr, p. 10) he was frightened by boys who threw cricket balls at him and he fled in fear, hiding behind a stand of trees. This memory haunted him and he resolved to do better. Later, at the age of eighteen, he was being chased by his cousin and brother and avoided capture by jumping off of a bridge. He fell a distance of twenty-nine feet and ruptured his kidney (Storr, p. 11). After three days he returned to consciousness and was unable to return to his studies for two months.

Just as he performed feats to attract the love and attention of his parents, he developed oratorical skills to overcome the impediments to his speech. He also risked his life and limb in dangerous enterprises in an effort to conquer his earlier fears. There was never a question of his native strength or energy. He lived to the age of ninety, surviving a heart attack, repeated bouts with pneumonia, strokes, and various surgical procedures. A lover of cigars, champagne, claret, port, and stilton, he denied himself very little. His personal work habits were legendary. The demons which haunted him were, at base, developmental.

He was also afflicted with depression, severe mood swings being part of his inheritance from the Churchills (Storr, p. 8). The first Duke of Marlborough was, in his later years, largely non-functional and much of the supervision of the building and decorating of Blenheim was done by his wife Sarah. For Winston, the depths of depression—particularly in his later years—were such that many have said it was a pity he did not die earlier, that, in effect, his final years were little more than a painful sentence.

These personal sufferings, however, were a crucial part of the man and the pattern of his experience. Rather than succumbing to them he strove to conquer them. The battles with his inner demons were transmuted into battles on behalf of the British people. Anthony Storr has analyzed this facet of Churchill's life and experience and his conclusions are so important that we quote them at length:

> Winston Churchill is still idolized, not only by those of us who remember his speeches in 1940, and who believe, as I do, that it was to his courage that we owe our escape from Nazi tyranny; but by men and women all over the world, to whom he has become a symbol, a personification of valor. But Churchill was also a human being, with the same needs, instincts, hopes, and fears which pertain to all of us. It is no disservice to a great man to draw attention to his humanity, nor to point out that, like other men, he had imperfections and flaws. Churchill, in spite of his aristocratic birth and social position, started life with disadvantages which he never wholly conquered, although his whole career was an effort to overcome them. Without these disadvantages he would have been a happier, more ordinary, better-balanced, and lesser human being. But had he been a stable and equable man, he could never have inspired the nation. In 1940, when all the odds were against Britain, a leader of sober judgment might well have concluded that we were finished. Political leaders are accustomed to dissimulation. Even when defeat at the polls is imminent, or the policies which they support have been shown to be futile, they will, until the eleventh hour, continue to issue messages of hope to their supporters. In 1940, any political leader might have tried to rally Britain with brave words, although his heart was full of despair. But only a man who had known and faced despair within himself could carry conviction at such a moment. Only a man who knew what it was to discern a gleam of hope in a hopeless situation, whose courage was beyond reason, and whose aggressive spirit burned at its fiercest when he was hemmed in and surrounded by enemies, could have given emotional reality to the words of defiance which rallied and sustained us in the menacing summer of 1940. Churchill was such a man: and it was because, all his life, he had conducted a battle with his own despair that he could convey to others that despair can be overcome.
>
> (pp. 4–5)

There are several points here to which we would draw attention. First, there is Storr's point that a less-troubled man would have been happier and more "ordinary." He would also, Storr argues, have been a lesser man. We do not believe that one can say that happy men (including men who are happy in "ordinary" things and ways) cannot be great. It is true, however, that our psychological models—including developmental models—imagine a continuum of normality or ordinariness and that while we are not urged to be ordinary we are often, at least implicitly, urged to be normal.

Second, Lawrence's esteemed poet, Yeats, said that all life seemed to him a preparation for something that never happens (Yeats, 1965, p. 91). Many men who have not experienced everyday forms of happiness and normality have turned elsewhere, some to evil pursuits, some to great accomplishments. The developmental lesson of an individual such as Churchill lies in the fact that many individuals feel as if their lives have not been normal. They have not developed along a prescribed path and they have, very often, suffered pain as a result. The experience of a man such as Churchill indicates that there is a relationship between pain and accomplishment, between developmental disability and personal triumph. If he succeeded in spite of—perhaps *because of*—his personal pain, it is possible that we could do so as well.

It is also clear that Churchill was inspired by great men, great deeds, and great words of the past. He was, as we indicated earlier, haunted by history—both the history of his nation and the history of his family. His pronouncements are often broad and contextual, locating current challenges and strife within larger patterns. At the age of thirteen (Manchester, p. ix), Churchill memorized the following passage from Macaulay's *Lays of Ancient Rome*:

> Then out spake brave Horatius,
> The Captain of the Gate:
> 'To every man upon this earth
> Death cometh soon or late.
> And how can man die better
> Than facing fearful odds,
> For the ashes of his fathers,
> And the temples of his gods?'

It is a passage that he would long remember, particularly when the Luftwaffe was bombing the cities of Westminster and London, laying waste to the church spires that Christopher Wren had built after the great fire of 1666 had done its damage. Churchill's military experience was quite real and his defenses, offenses, and counter offenses carefully and skillfully planned, but in the back of his mind was history, memory, and poetry—patterns and representations of experience that served to inspire him in dark moments.

In a speech before the House of Commons in 1940 (Manchester, p. iv), he said, "History with its flickering lamp stumbles along the trail of the past, trying to reconstruct its scenes, to revive its echoes, and kindle with pale gleams the passion of former days." Churchill, like T. E. Lawrence, followed that lamp and listened for those echoes. In some respects it is a metaphor for his own life: seeking meaning and inspiration for the present in the lessons and stories of the past.

As we noted earlier, Lawrence served for a time as a member of Churchill's staff. In a series of accounts of his famous contemporaries (1990), Churchill recorded his memories and impressions of Lawrence. It is interesting that Churchill—who relied upon Lawrence's technical advice and expertise—saw him in terms that can only be described as romantic. He succumbed, in a sense, to the artistic construction, though there was clearly a good deal of reality undergirding it:

> From amid the flowing draperies his noble features, his perfectly chiselled lips and flashing eyes loaded with fire and comprehension shone forth. He looked what he was, one of Nature's greatest princes . . . . Whether he wore the prosaic clothes of English daily life or afterwards in the uniform of an Air Force mechanic, I always saw him henceforward as he appears in Augustus John's brilliant pencil sketch.
>
> (p. 98)

> In principle the structure of the story is simple. The Turkish armies operating against Egypt depended upon the desert railway. This slender steel track ran through hundreds of miles of blistering desert. If it were permanently cut the Turkish armies must perish; the ruin of Turkey must follow, and with it the downfall of the mighty Teutonic power which hurled its hate from ten thousand cannons on the plains of Flanders. Here was the Achilles' heel, and it was upon this that this man in his twenties directed his audacious, desperate, romantic assaults.
>
> (p. 102)

But there is more to Lawrence than the accomplishments and the image. Underneath is something of a different order. Churchill comments on the fact that Lawrence the charismatic leader was also capable of working as a member of a staff, exemplifying the dedication and fidelity that such a task demanded:

> I always received united advice from two or three of the very best men it has ever been my fortune to work with. It would not be just to assign the whole credit for the success which the new policy secured to Lawrence alone. The wonder was

> that he was able to sink his personality, to bend his imperious will and pool his knowledge in the common stock. Here is one of the proofs of the greatness of his character and the versatility of his genius.
>
> (p. 101)

The end of Churchill's reminiscence is more a peroration than a conclusion. He obviously feels the powerful sense of Lawrence's absence, the realization of the enormity of the role that Lawrence might have played (had he not died accidentally in 1935) in the struggle against the Nazis. The passage is too long to be quoted in full, but it is sufficiently important to be quoted at length:

> In these days dangers and difficulties gather upon Britain and her Empire, and we are also conscious of a lack of outstanding figures with which to overcome them. Here was a man in whom there existed not only an immense capacity for service, but that touch of genius which everyone recognizes and no one can define . . . . He always reigned over those with whom he came in contact. They felt themselves in the presence of an extraordinary being, felt that his latent reserves of force and will-power were beyond measurement. If he roused himself to action, who should say what crisis he could not surmount or quell? If things were going very badly, how glad one would be to see him come round the corner.
>
> Part of the secret of this stimulating ascendancy lay of course in his disdain for most of the prizes, the pleasures and comforts of life. The world naturally looks with some awe upon a man who appears unconcernedly indifferent to home, money, comfort, rank, or even power and fame. The world feels, not without a certain apprehension, that here is someone outside its jurisdiction, someone before whom its allurements may be spread in vain, someone strangely enfranchised, untamed, untrammelled by convention, moving independently of the ordinary currents of human action . . . . Lawrence was one of those beings whose pace of life was faster and more intense than the ordinary. Just as an aeroplane only flies by its speed and pressure against the air, so he flew best and easiest in the hurricane. He was not in complete harmony with the normal. The fury of the Great War raised the pitch of life to the Lawrence standard. The multitudes were swept forward until their pace was the same as his. In this heroic period he found himself in perfect relation both to men and events.
>
> (pp. 103–104)

What Churchill has seized upon, of course, is the extent to which Lawrence was not only *different*, out of the *ordinary*, but, simultaneously, in possession of virtues which overshadowed those of ordinary men and made him an inspiration to them. Churchill perceives this and meditates upon it at such length because in Lawrence he sees something of himself.

Churchill too was anything but an ordinary man though he was denied some of the comforts enjoyed by ordinary men. He could carry the burdens of his country but could not enjoy the attention and love of his mother and father or the respect of his schoolmates. After his country was saved he was rejected by the electorate. If his spirit was never fully whole, he attained his heroism in the same manner that Lawrence attained his. His life and the actions of that life remained grounded in virtue. He never gave up hope. He always exhibited willpower, purpose, and competence. He was faithful and loving, caring, and exceedingly wise. He was loved by his family and despite an occasional attempt to challenge his accomplishments he continues to be revered by the nation for which he is a continuing *exemplum.*

Setting personal standards, acting, and even acting out, but if things were going very badly, how glad one would be to see him come round the corner.

Of all of the heroic examples that we adduce in this book, General George S. Patton, Jr. is likely to be the most controversial. "Old Blood and Guts" ("Yes—our blood, his guts," a soldier in the film *Patton* [1970] says) is a figure associated with a military posture that could be characterized as aggressive, fierce, and relentless. Patton may have helped to defeat the Nazis but, in Sicily, he also slapped two American GI's who had been hospitalized for combat exhaustion, in the process stupidly jeopardizing both his military career and his future reputation.

Those who served with him whom we have interviewed often speak of him with an "I could tell stories" look in their eyes. The response is never a simple, "Yes, he was a very great man, the finest I have ever known." However, whether these individuals liked the man or not it is always clear that he commanded their respect. A strict disciplinarian with an eye for detail in all things—a pattern of activity which characterized his behavior at VMI and at West Point, long before his appointment as a general officer—he is often remembered for expecting and requiring odd things.

He is also remembered for achieving dramatic results. One should say at the outset that he commanded the respect of the Nazis as well. He defeated them at El Guettar, Tunisia and he defeated them in the Battle of the Bulge, saving the 101st Airborne, trapped at Bastogne, in the process. He may not be the man everyone would consider a hero, but as Churchill said of T. E. Lawrence, if you found yourself in great trouble, surrounded, for example, by multiple divisions of the Wehrmacht, it would be a source of immediate comfort to see his face coming around the corner to your rescue.

Prior to the Normandy invasion the allies mounted a great propaganda offensive in which they created fictitious units, disinformational radio traffic, and even artificial weapons systems (inflatable tanks, for example) that would be misinterpreted by aerial photographers. The point of this was to lead the Germans to believe that the allied invasion would actually come at Calais rather than at Normandy. The putative commander of this operation was Patton—the individual whom the Nazis would logically expect to be the leader of an action of such magnitude and importance.

His life and personality are rich, complex, and fascinating. The story begins and ends with his own family, which was so important a part of Patton's life. In the first place, he was extremely wealthy. He had no need to pursue a military career. (His hand-tailored uniforms and embassy-size Washington residences were a source of surprise to those who did not know his family background.) His maternal grandfather was Benjamin Davis Wilson, a legendary figure in the history of southern California.

Mount Wilson, which rises among the San Gabriels above Pasadena, is named for him, commemorating the fact that he blazed the trail up the mountain in search of timber for his wine and brandy casks. He was, by 1871, the largest manufacturer of wine and brandy in the state of California. A former trapper, Indian trader, and storekeeper, he was elected Alcalde of the San Bernardino district. He was also elected county clerk, city councilman, and first mayor of the city of Los Angeles.

Known as "Don Benito," Wilson was also, to say the least, a significant owner of land. His holdings eventually numbered approximately 14,000 acres, spreading south from the San Gabriel foothills and comprising what would now include Pasadena, South Pasadena, San Marino, Alhambra, and San Gabriel. He owned orchards and vineyards in downtown Los Angeles on the current site of the city's Union Station. He also owned acreage to the west, including the present site of the University of California, Los Angeles, in Westwood.

This tradition of local leadership and responsibility continued with Patton's father, who was elected first mayor of the city of San Marino, a wealthy and beautiful area between Pasadena and San Gabriel. The last Patton family home sits just beneath the lush grounds of the Huntington Library, near Lacy Park, the latter a gift of the family to the community. The Pattons worshipped at the Church of the Saviour in San Gabriel and they are remembered there in a very striking way. Not only is there a statue of Georgie (as he was called by friends and family) in the church garden; there is also a stained glass window—in the church proper—depicting Patton in the cupola of his tank next to a representation of Saint George, slaying an earlier dragon than the dragon that was the Third Reich. The pictures are accompanied with biblical quotations on the fighting of the good fight. This mix of religion, war, ancient chivalry, and poetry is a very natural one for Patton, given his experience and given his view of the world.

His father's side of the family is equally interesting. His first American forebear, Robert Patton, married the daughter of Hugh Mercer, a physician on the Jacobite side at the battle of Culloden. In the Glorious Revolution of 1688, the Stuart King, James II, lost his crown to William of Orange and James's daughter—William's wife—Mary. The Stuarts moved to Catholic France but still claimed the right to the English throne. Their opponents called the Stuart "king" in exile across the waters, the "Pretender," while the supporters of the Stuarts were termed *Jacobites* after the Latin word for James, *Jacobus*. Successive (but not successful) Pretenders rose both in 1715 and in 1745. Prince Charles Edward Stuart ("Bonnie Prince Charlie") came in 1745 to reclaim his throne; he gathered a cohort of Scottish troops and proceeded to invade England.

The invasion was, for a time, a considerable success. Many of the current roads and canals of England were later built in response to the realization that the Pretender and his troops were able to get as far as they got because of the lack of transportational infrastructure that would have been required to oppose them.

The Pretender and his forces drove as far south as Derby, but they were then driven north, practically to the coast of Scotland. Upon Culloden moor, a now-melancholy patch of ground a few miles outside the city of Inverness, Charles Edward Stuart's forces were slaughtered by those of the Duke of Cumberland. The prince survived by dressing as a woman and hiding on the Isle of Skye under the protection of a woman named Flora Macdonald (who is now memorialized by a monument at Inverness Castle).

After the victory the English established policies and conducted operations designed to keep such an invasion from ever occurring again. They banned the carrying of weapons in Scotland; they banned the wearing of tartans, and they set about the systematic destruction of every military-aged man they could find. Thus, the battle of Culloden is both a divisive and defining event in English and Scottish history. Individuals continue to visit the battlefield, throwing pennies into collection vessels with tears streaming down their faces.

The liqueur Drambuie (reputed to be Prince Charles Edward's favorite drink) includes references on each bottle linking the drink to the '45 (as it was termed, though the battle of Culloden occurred in 1746). In short, the battle and the war are the stuff of legend and memory and Patton's ancestor, the surgeon Hugh Mercer, was there. Mercer was also a friend of George Washington's and a Brigadier General in the patriot army, dying a military hero after falling in the battle of Princeton.

Robert Patton—Mercer's son-in-law—was a Scotsman whose real name he would never reveal (even though his family begged him as he lay dying).

He assumed the name of Patton for reasons described by Robert H. Patton, Georgie's grandson:

> Back in Scotland [Robert Patton] was a wanted criminal. He'd fled the country in the mid-1760's with a price of five hundred pounds on his head, changing his name to Robert Patton to better lose himself in America. The name Patton was a popular alias of Scots emigrants. In Gaelic it means king's pensioner, *pensioner* in its original usage being a gentleman-at-arms. Taking the name was a declaration of service to the king. The *true* king, that is, who in the opinion of many in the southern colonies was not the king of England but Charles Stuart.
>
> (1994, p. 13)

Thus, we do not know Georgie Patton's actual surname. What we know is that his paternal ancestor was a Scotsman who allied himself, through the choice of his name, with the Jacobite cause. He was also a man of sufficient importance to warrant a £500 bounty, a tidy sum of money in the eighteenth century, comparable in modern currency to, perhaps, $100,000. Eighteenth-century rural clergymen, living modestly, could support themselves and their families for a year on £40 or less.

Thus, the Patton family tree is one enshrouded in mystery, but one associated with battles, legends, opposition to authority, and defense of the "true" monarch. It is also associated with military accomplishment and public acclaim. Hugh Mercer's funeral in Philadelphia in 1777 was attended by 30,000 mourners and several paintings were done of Mercer, representing his service at the battle of Princeton. The most famous of these paintings—John Trumbull's—hangs today at Yale University (Patton, p. 17).

In ensuing generations this pattern continued. Georgie's great uncles, for example, fought in the Civil War. One of his uncles, Walter Tazewell Patton, was killed at Gettysburg during Pickett's charge. Georgie's grandfather, George Smith Patton, was colonel of the 22nd Regiment under Jubal Early. He was mortally wounded at the third battle of Winchester in September, 1864. His saddle—stained with his blood—stayed in the family; Georgie learned to ride on it.

He generally associated himself with the Virginia, rather than the California, Pattons. These individuals were upper-class, educated, civilized, and martial. During his frequent moments of self-doubt he thought of his heritage and genealogy, assuring himself that the quality of his ancestry would serve as a support and carry him through difficult times.

Georgie's strongest subject in school was History and he characteristically saw things in historical terms. His father conveyed to him a view of the world that was nearly mystical and we know that he was personally interested in telepathy and reincarnation—a point that is made in the Franklin Schaffner film in a scene in which Patton describes a Carthaginian/Roman battlefield and battle to General Omar Bradley and then says, "I was here, two thousand years ago." His grandmother had claimed that she possessed what the Scots term "second sight." She told people, for example, that she "knew" whenever her husband was wounded in the Civil War. Whatever one thinks of reincarnation and telepathic communication, it is clear that Patton felt the pull of the past and the influence of the past through his family. He lived in fear of shaming his ancestors and he felt their presence, both spiritually and directly through such mementos as his grandfather's bloody saddle.

The men in his family traditionally attended the Virginia Military Institute, his grandfather graduating in 1852, his father serving as first captain during his time there a generation later. When Georgie arrived there himself and was measured for his uniform he discovered that he wore precisely the same size as his father and grandfather, a source of encouragement and comfort to him (Blumenson, 1985, p. 43). This was what he needed at a desperate time: a good omen.

All through his youthful life his father had read to him, inculcating attitudes and wisdom. He particularly read Sir Walter Scott, Homer, Xenophon, Shakespeare, Kipling, and the Old Testament (Blumenson, p. 29). This is an interesting list, including, as it does, works that often concern the intersection between history, legend, and military actions, all matters of consequence to Georgie, both in his youth and throughout his life.

His father read to him for a specific reason. George S. Patton, Jr. was dyslexic and had great difficulty both in reading and in writing. In part because of the impact of the academy award-winning film, Patton is often associated with salty oratory that can be strikingly eloquent. He was a poet, for example, as well as a story teller. The film begins, in fact, with a long, stylized speech to his troops. He was, of course, an interesting speaker with a flair for quotation. In part that was because he memorized lengthy passages from literary works and developed his abilities to recite them in detail. He did this, quite simply, because his dyslexia forced him to develop aural and oral skills.

He was not, by standard measures, a good student. On the contrary, his learning disability was a severe challenge to him both intellectually and spiritually. He would often become frustrated with himself and describe himself as stupid. He was self-deprecating and insecure, frequently describing himself to his future wife and to his parents as *worthless*. After his plebe year at West Point (to which he went after studying at VMI) his academic record was examined and a decision was made to "turn him back," i.e. to force him to repeat his first year because of his failure in mathematics.

Fortunately, he had the enduring support of his family and, after his marriage, of his wife, Beatrice. In many ways, she picked up where his family left off, consoling him, encouraging him, and urging him on. The briefest inspection of the details of his life and the depiction of that life in letters reveals how different the real man was from the public image of the man that was commonly held. The fierce and bloody warrior, ready and anxious to engage the enemy in mortal combat, was frightened, frustrated, insecure, and extremely dependent.

He was so homesick and dependent, in fact, that his mother and aunt often resided near him both while he was at VMI and while he was at West Point. He also received frequent visits from his sister. They were there to comfort him, to encourage him, to look after him, to reassure him, and to provide the continuity with the family life which he so sorely missed.

Georgie was not developmentally disabled in the sense that he lacked the love of his parents and family. On the contrary, he was disabled by his dyslexia and its accompanying dependency; his father helped him to compensate for his difficulties by reading to him and teaching him. However, the love and attention which he received was so abundant that he came to require it. His family loved and helped him, but the obligations imposed upon him by family legacies and traditions haunted him. He desperately needed to meet the *family's* expectations—the family being his entire heritage, not just the nuclear family. His nuclear family, his aunt, and his wife Beatrice helped him in this regard, but the dependency which resulted was of an order that one might fairly term abnormal. It is, in short, not common for VMI and West Point cadets to have their families in virtual residence nearby so that they might be regularly comforted and nurtured.

His feelings of personal inadequacy were more than intellectual concerns, brought on by his dyslexia. He also feared that he might be a coward. When faced with challenges, he experienced (as many do) shortness of breath, sweaty palms, and sleeplessness. He also suffered from such physical maladies as hay fever and hives.

Thus, he sometimes tested his personal courage in foolhardy ways. Once at the West Point rifle range, for example, he was in the pits, pulling and pasting targets and recording scores. Overcome by the sense that he might not be able to face enemy fire, he stood up—at the level of the range—with bullets whizzing around him. He thus proved to himself that he could stand amid mortal fire without showing fear.

To the unbiased observer this was colossally stupid, but to him it was quite necessary. There were other such events in his life where he felt it necessary to demonstrate his freedom from fear. In one of his classes, for example, the cadets were studying electricity and Georgie volunteered to absorb a strong shock in order to demonstrate his courage and his ability to survive (Blumenson, p. 55). Actions such as these recall the widely-reported experiences of

G. Gordon Liddy (1980), who conquered his fears by facing them directly: eating a rat to allay his fears of rats, climbing a tree in a thunderstorm to allay his fears of lightning, etc. To the average observer this is not only somewhat bizarre, it is, perhaps, pathological.

As we indicated earlier, Patton is a far more controversial hero than most. The point, we believe, is not that his actions were often strange or out of the ordinary, though many clearly were. The more interesting point is that his personal development was both positively and adversely affected by his family history, just as the love of his parents, friends, and family provided a necessary foundation for his success even as they created a level of dependency that most would consider excessive.

In a sense his entire life was a succession of contradictions. There is a particularly telling scene in the film *Patton*. It occurs during the Battle of the Bulge. Patton is at his headquarters, a large, cavernous structure in which his voice reverberates. He is planning for imminent battle and in one heated moment he cries out to the members of his staff, "If we are not victorious, let no man come back alive." His aide, Colonel Charles Codman, is shaken and tells the general that he should be more guarded, that his men do not know when he is serious and when he is not. "It's not important that they know," Patton responds. "It is only important that I know."

Throughout his military career Patton was an actor. By nature he was kind, emotional, and deeply sensitive. Unsure of himself, he was naturally reticent and responded immediately to objects of beauty, be they natural or man-made (Blumenson, p. 12). He feared, however, that these characteristics would stand in the way of his success as a military officer. Thus, he suppressed them as best he could, cultivating, in their place, their opposites.

Thus, he appeared gruff and choleric, scowling often and emitting hard words and cold stares. He cultivated his appearance as an actor might, staring at himself in the mirror and adjusting his expressions. He worried about his bearing and posture; he was scrupulous with regard to his appearance and, also, scrupulous with regard to the appearance of his men.

One of the prerogatives of general officers is the latitude to vary their uniform from the common mold. (Patton, in fact, even designed a uniform that would have been worn by Armor troops—a flashy but efficient suit, covered in buttons and topped with a round football helmet.) His own accessories (or affectations) are well-known, particularly his carrying ivory-handled pistols. Only a skilled actor and a convincing leader could carry off something like that.

We also know that the Pattons were fond of masquerades and costume balls; photographs exist, for example, of Patton dressed as Rhett Butler, with Bea as Scarlett O'Hara. He also appeared as King Arthur to Bea's Guinevere, mounted upon a dark horse, led by a harlequin figure and trumpeter, both in full costume. In his youth he tried on identities and roles and his fantasies were fed by the books which his father read to him.

The general as actor is, of course, a dangerous role, given the responsibilities involved. With Patton, however, the role was combined with courage, skill, and dedication. He could, in short, play the part as well as look the part, but it was very much a role. The real Patton—the suppressed Patton—came through with some frequency: Patton the poet, Patton the family man, Patton the sensitive man subject to moments of heart-felt words and streaming tears.

We would not describe this role-playing as neurotic. Part of his natural character involved a great desire for military success and personal fame. He was not persuaded to be such a man or forced to be such a man. His family was supportive, not tyrannical. Patton himself sought to be a part of his family's grand tradition. He knew the identity he was seeking to attain. Thus, his actions in assuming his role are quite conscious ones. They are not the result of some deep-seated, misunderstood conflict, with Georgie being pulled in two directions and torn apart in the process. He knew what he was and he knew what he wanted to be and he realized that he would have to make behavioral and imagistic changes in his life and appearance if he were to succeed in reaching his goal. He resolved to do it and he did it.

It did not always come easy. In moments of great stress and danger he was buoyed by the traditions into which he had been born. He heard, as it were, the voices of his ancestors, pulling him on. He responded to those voices. While his actions may have been accompanied by a certain amount of bravado, there is no question of his competence and courage. After his first major engagement in World War I, where he was wounded and continued to direct his troops though the blood was streaming from his body, he received the Distinguished Service Cross, the second highest medal for valor after the Medal of Honor. Simultaneously, for his training of the first group of American tankers he received the Distinguished Service Medal, an award for conspicuously successful performance.

He was also immensely lucky. Selected as General John Pershing's aide for his operations in Mexico, hunting Pancho Villa, he was later selected to shape what would become the Army's Armor branch. He was, literally, the first Armor soldier and commander in the U.S. Army, learning from the English and French and developing battle doctrine for the use of tanks in World War I. While in Mexico he was assigned the task of hunting one of Villa's officers. He and his comrades conducted the operation using motorized vehicles. The operation was a success and it was, in fact, the first motorized operation in the history of the United States Army.

Thus, when Patton is portrayed as a former horse cavalryman leading the Army into the age of tanks and armored vehicles, the characterization is absolutely accurate. Later, of course, he would establish a base in the Mojave Desert for the training of armored troops in preparation for operations in North Africa. That site, just east of Palm Springs, is now a museum honoring Patton and his many accomplishments.

His indomitable will is also revealed in the fact that he was a part of the American team in the 1912 Olympics in Stockholm, participating in the modern pentathlon. He achieved fifth place in those games and he was the only fencer to beat the French champion. This is a characteristic accomplishment, combining as it does both physical challenge, personal dedication and competence, and the mystique of chivalry and ancient weaponry.

Patton's end was of a different order. Severely injured in a vehicle accident that occurred after the war, he died a lingering death, protracted by his great physical strength and endurance. Lying in bed with a broken neck, his damaged vertebrae resulted in full paralysis below the neck. Eventually the physicians went so far as to place him in traction with weights suspended from lines attached to hooks that were inserted under his cheekbones. He died of pulmonary edema and congestive heart failure—just like Lawrence, a warrior dying as a result of a freak motorcycle accident in which no one else was hurt.

Controversy followed him to the end of his career, as he employed former Nazis to aid in the administration of a conquered Germany. His goals were pragmatic (to keep the industries operative), his political sensitivities minimal. Nevertheless, the tribute paid to German efficiency implicit in those actions was more than counterbalanced by the violence visited upon the Third Reich by the Third Army.

The true Patton is a complex figure: an inspiring warrior who himself needed inspiration and consolation. In a letter to his wife from France during World War I, he wrote:

> I certainly don't see any stars in prospect for me but one can always try. Sometimes I think I don't try as hard as I ought but probably I do . . . . I would give a lot to have you consol [*sic*] me and tell me that I amounted to a lot even when I know I don't.
>
> (Blumenson, p. 95)

This is the real Patton—writing in halting English, reaching out for love and help, expressing his dependence and self-doubt, concerned about his career and future but expressing a resolution to do his best. This is the man accompanied to VMI and West Point by his mother and aunt and whose mother and sister were always nearby to hold his hand. This is the same man who heard the voices of his ancestors, was promoted to the rank of full colonel at the age of 32, and who received the Distinguished Service Cross as a result of his actions in his first significant engagement. Like his Civil War ancestors, he led from the front, not from the rear. He achieved his goals despite the challenges which he faced and he helped—significantly, one must say—to defeat the German war machine.

He did so despite cognitive and developmental disabilities. What he did not lack, however, was virtue. He had hope, willpower, purpose, competence, fidelity, love, care, and wisdom. Wisdom, in particular, was one of his mainstays, for his deep knowledge of history and his deep awareness of military doctrine and technique show themselves in his words and actions. Most of all he possessed courage, a special virtue for the military figure. There is a saying in the Army that courage is fear directed at the enemy. That is an apt description of Patton and his ways. He took his shortcomings and transmuted them into strengths, with the most dramatic of results.

In addition to historical examples we have many fictional characters who are an important part of our cultural heritage and whose wounds and abnormalities have led to heroism.

Enduring elements of popular culture often speak to a society's deepest beliefs. For many, the notion of heroism is encapsulated by the exploits of comic superheroes. Certainly they have had an important impact on our cultural history and the extent of their resonance is indicated by the fact that their stories have appeared in multiple media. There have been, for example, Batman serials, television programs, feature films, and cartoons as well as comic books. Batman has recently even appeared as the protagonist in an Andrew Vachss novel, dealing with child abuse. The superheroes are staples of Saturday morning cartoon programming and they often appear in the form of so-called "action figures," a more masculine term for plastic or rubber dolls.

What is striking about the superheroes as a group is the fact that while they all exhibit virtue in the cause of justice, truth, and righteousness, they also exhibit developmental disabilities. Batman (Bruce Wayne), for example, is an orphan who witnessed the murder of his parents. His ward, Dick Grayson, who participates in his efforts to eradicate crime as Robin is also an orphan.

Superman (Clark Kent) lives in exile from his true home on the former planet, Krypton, the destruction of which annihilated its citizens and orphaned him. While his feats of strength and speed are useful in enabling him to combat crime, they also set him apart from his schoolmates. In effect he is forced to conceal his true abilities. He remains alone, apart, different, anything but "normal." Like Batman, Zorro, the Lone Ranger (the sole survivor of a massacre) and, to some extent, the Phantom, his life consists of the systematic concealment of his personal identity.

Other superheroes are frequently the products of some scientific or industrial accident and their special abilities are also, to some degree, a burden (Spiderman, the Incredible Hulk). At times they are hideously misshapen (Swamp Thing), nearly destroyed (Robocop), or, to say the least, "differently abled" (Wolverine or The Thing).

These figures have endured in our culture far longer than actual human beings such as the singing cowboys (Roy Rogers and Gene Autry, for example) who were skilled musicians, handsome, well-dressed and successful with women. They have also lasted much longer than the developmentally woundless serial heroes of the 1930's and 1940's, Flash Gordon, Buck Rogers, and Don Winslow of the Navy (and of the Coast Guard).

Our contemporary popular culture heroes are anything but "normal" and their disabilities and different abilities are generally central parts of their personality. They share such virtues as courage and tenacity with their forebears, but their lives have often been disrupted, their personal growth impeded, their personal lives nonexistent. Spiderman is an exception, since his marriage, but he has no children and his overwhelming need to utilize his abilities in the interest of justice continually disrupts his marriage with M. J. and complicates their life together.

For a time we were awash in vigilante movies or, at least, films that suggested that the judicial process as we knew it was not bringing justice as frequently as it should. The success of many of the Clint Eastwood movies was attributable to the temper of the times as well as to the direction of Sergio Leone and Don Siegel. In that regard, Harry Callahan lost his wife as well as most of his partners, but he never gave up his essential virtues. A similar pattern of personal loss occurs in a number of the Eastwood films and serves as the basis for their plots (*High Plains Drifter*, *The Outlaw Josey Wales*, *Unforgiven*, for example). We know nothing about the past of Leone's hero—the man with no name. All we know, as Churchill said of Lawrence, is that if we were in trouble we would be very happy to see him come around the corner.

In his essay "The Decay of Lying," Oscar Wilde argued that life imitates art far more than art imitates life. The fog that we see in London was first imagined for us by Dickens, just as the fields of flowers we now see in France were first rendered by Monet. Aristotle reversed the relationship. Art is an imitation of an action, not a direct imitation of life, but an imagined action that exists in an analogous relation to life. Art exaggerates and isolates and highlights so that we can see things more clearly. Dryden once said that statues must be larger than life. If they were not, they would look smaller than life to those of us who stand at their feet.

However one conceives of it, the relationship between art and life is a close one, particularly the relationship between our daily lives and the popular culture which we consume. New York Publishers and Hollywood producers continually search for novels and scripts that bear on pressing, present-day concerns. One of the reasons that genre fiction has often been successful is that it can succeed in representing (or in purporting to represent) aspects of an entire age (as in regency romances). Similarly, it can depict an important aspect of our current experience in such a way that we gain an appreciation for the dimensions of that phenomenon or experience through the consumption of a six-hour novel or ninety-minute film.

Two recent novels reinforce this pattern. Mark Haddon's *The Curious Incident of the Dog in the Night-Time* features a youthful protagonist named Christopher Boone; Jonathan Lethem's *Motherless Brooklyn* focuses upon a young man named Lionel Essrog. Christopher suffers from borderline autism, while Lionel is afflicted with Tourette's syndrome. Each is presented with a crime that each is compelled to solve, even though the solution of the crime will take each one of them out of his personal comfort zone and will do so literally—forcing each to leave his by-now comfortable physical surroundings and travel into the *terra incognita* of, respectively, London and Maine. In each case, the protagonist's affliction forces him to bravely face down his fears, while the nature of each affliction is such that it actually helps prepare him to solve the case. The 'wound' that each protagonist carries is multi-faceted and while each 'wound' marks its possessor as 'different' developmentally, it contains elements which makes the possessor more effective as he pursues his quest and, ultimately, heroic as he achieves his success.

Both of the books were great commercial successes, particularly Haddon's, and *Motherless Brooklyn* is being made into a film, featuring Edward Norton as Lionel. In each case, the writing is expert, but both Haddon and Lethem are also successful because of the resonance of their central theme—the triumph of the developmentally-challenged through the leveraging of the particular elements of personality initially perceived to be deficits rather than, ultimately, strengths.

We now expect our popular entertainments to be scrupulously accurate with regard to detail. Just as the social melodrama of the sort written by Arthur Hailey represents the innerworkings of an important industry, the contemporary crime novel or crime film represents aspects of experience in which large numbers have a particular interest, such as casino gambling, computer crime, crack dealing, or counterfeiting. We read about such activities in the newspaper or have some personal, peripheral experience with them, but the film or novel enables us to see aspects of such an activity in detail and it forges a personal relationship between us and the activity by creating vivid characters who evoke a strong personal response from us. In the case of the film, we read reviews in advance and discuss the film after its screening with our fellow filmgoers. In the case of the novel we internalize and personalize the experience as we hear the narrator's words in our head, spoken in our own voice.

The degree to which current trends in popular culture mirror the concerns of this study is striking. Clearly we are interested in developmental issues, particularly developmental issues which involve deviation from patterns considered to be normal. One of the most popular recent subjects for fiction and film writing has been the serial killer. Indeed there has been a spate of such works and each time a publisher or producer says that he will buy no more we suddenly see yet another example appcar. Earlier we referred to the work of Robert Ressler, the former profiler at the Federal Bureau of Investigation, and noted the fact that, in Ressler's studies, many serial killers suffered developmental disabilities which were not balanced by the virtues which we saw in our heroic models. Both had disabilities, but the virtues of the hero led to a life which others now esteem, while the disabilities of the serial killer (unredeemed by virtue) led to a life of horrid crime.

The most successful of the serial killer novels/films has been Thomas Harris/Jonathan Demme's *The Silence of the Lambs*. Harris is the acknowledged master of the psychological thriller/crime novel, with three other successes in addition to *Silence*: *Black Sunday*, *Hannibal,* and *Red Dragon*, the latter initially filmed as *Manhunter*, directed by Michael Mann. *Silence* won five Academy Awards, including the award for best picture of 1991. Recently, Harris has added a prequel: *Hannibal Rising* (2006).

The extent to which this story is now an important part of contemporary popular culture suggests two things: we are very interested in human development and we are very interested in the manner in which developmental disability can be transcended through the practice of virtue.

Indeed, *The Silence of the Lambs* and the sequel *Hannibal* are part of a larger contemporary pattern, the spate of films, shows and television programs built upon the Beauty and the Beast tale. In addition to the television program featuring Vincent and Catherine (1987–1990) and the Disney animated feature (1991), we have seen Andrew Lloyd Weber's amazingly successful adaptation of the Phantom of the Opera story and such interesting variations on the theme as the film *Darkman* and *Terminator II* (the *Terminator* film now transmuted into a television series).

The Beauty and the Beast tale is a story of human development in the face of personal suffering. The most famous version of the tale was first published in 1756 by Madame Le Prince de Beaumont; cinematically it was most faithfully reproduced by Jean Cocteau (1946). Among fairy tales, this story falls within the broad category of animal-groom tales, in which humans marry (what at first appear to them to be) animals.

The story tells of a merchant with three daughters, one of whom is particularly lovely. He loses his wealth and sets sail from home in an attempt to recoup his losses and rebuild his financial life. Two of his daughters request luxurious gifts from him upon his return; Beauty, on the other hand, tells him that she would like nothing more than a single rose.

The journey is a failure. During the return voyage the merchant becomes lost in a storm. He discovers a magic palace and steals a rose from its garden. This enrages the inhabitant of the palace—the beast—who demands in return either the life of the merchant or the hand of one of his daughters.

Beauty saves her father by volunteering to be his substitute, but once inside the palace she becomes its mistress and develops respect for the beast who is its master. Missing her sick father, she returns home but stays too long. Then, missing the beast, she returns to the palace to find him grieving and at the brink of death. When she announces her love for him, he is turned into a prince and the two of them marry.

The success of the story over the last two and a half centuries turns on the fact that it deals with such basic issues as the love between men and women, the need for young adults to leave home and create new lives and families, and the fact that that process is not always easy. Daughters are not anxious to leave their fathers and the young men who court them may at first appear to be beasts. The rose, a traditional symbol of virginity, stands at the center of the story just as it stands at the center of the developmental process, involving, as it does, knowledge, experience, love, commitment, and a step into the unknown.

The film *Terminator II* parallels but inverts this pattern. It is built upon a mother-son rather than father-daughter relationship and modernizes the conflict by treating the son's search for paternal rather than marital love. At a time when public attention is increasingly focused on absentee fathers, dysfunctional families, and domestic violence, the *Terminator II* cyborg functions as a singular alternative: the father who is absolutely dependable, even down to the last and most final detail—his willingness to give up his life in order to protect his son. Like a number of the comic book superheroes the cyborg is 'disabled' in the extreme (he is not human but a machine), yet he exhibits virtue in a sufficiently powerful way to evoke a strong human response from the audience.

The most interesting 'beast' to charm a 'beauty' may well be Dr. Hannibal Lecter, the psychotic psychiatrist who first appears in Thomas Harris' *Red Dragon* and who returns in Harris' best-selling *The Silence of the Lambs* (and, most recently, in the sequel *Hannibal* and prequel, *Hannibal Rising*).

In the book as well as in the Jonathan Demme film, the imprisoned Lecter is sought out by the FBI for assistance in tracking a serial killer. Beastlike indeed, Lecter (nicknamed Hannibal the Cannibal) bites and chews and swallows with more avidity than Disney's Beast and with far greater frequency than the docile and talky television beast, Vincent. At the same time, he is (like Vincent) courtly in his approach to the young FBI cadet who is sent to interview him.

In one of the most striking moments of the film, Lecter hands the cadet (Clarice Starling) the serial killer's file through the bars of his cage. When she takes it he touches one of her fingers in a gesture that speaks volumes about the distance implicit in their relationship. While they are technically opponents, they work together and exhibit mutual respect.

After the serial killer has been caught and Lecter has escaped, he calls Clarice on the day that she graduates from the FBI Academy. He promises he will not seek her out, since he believes the world to be a more interesting place with her in it. He has had his single touch, more than the usual courtly lover ever obtains.

The Silence of the Lambs plays off the key notion in the beauty and beast pattern: the relationship between a young woman and her father. There are generally no mothers in these stories. Clarice Starling has become an FBI cadet out of loyalty to her father's memory—a marshal killed years earlier in the line of duty. Her career has been advanced by Jack Crawford, chief of the behavioral sciences unit at Quantico, and, of course, by Lecter: three men, all mentors, all very different, but sharing a common concern for a woman who aspires to be a professional in a world that most men could not even face.

In each case we see heroic qualities attached to personal psychological disabilities. Clarice is an orphan; Crawford loses his wife Bella in the course of the novel (this is not treated in Demme's film), and Lecter (we learn in *Hannibal*) has lost his beloved sister under hideous circumstances.

Clarice is an actual heroine, breaking and solving the case even though she is a relatively inexperienced cadet. Crawford is the traditional American hero—short on words but honest, straightforward, and true. Lecter is an odd choice for heroism, but he exhibits a number of virtues, treats Clarice with respect and, eventually, love, and—both in *Silence* and in *Hannibal*—triumphs over circumstances through tenacity, will, intelligence, and courage. Like many of the great monsters, he has his own charms and exhibits the tastes, education, and intelligence of a figure such as Dr. Henry Jekyll; he simply requires us to forget the Mr. Hyde side. In *Hannibal* he masquerades as one "Dr. Fell" in a museum in Florence. *Fell* the adjective means *savage*, as the noun means *skin* or *hide*, so that the connection with Jekyll and *Hyde* is made quite explicit.

Why should the beauty and beast story be so resonant now? Why has it appeared in so many different forms? Certainly it has much to say to those involved in contemporary relationships. In an era of safe sex and AIDS, for example, there are more distant, courtly lovers at large in our stories than have been sighted since the Petrarchan sonnet sequences of the renaissance. In an age aspiring to more effective, more caring communication, sex is often being replaced by talk.

In a time when women have taken their rightful place within the power structures of society, a troglodyte like Disney's Gaston—the beast's rival in the animated film—becomes an instant object of opprobrium when he promises Belle a life in the same village forever, with a house full of little Gastons. We are introduced to him when he kills a duck in midflight, a bit of violence that is uncommon in Disney, particularly when it does not involve a major plot point (such as the death of Bambi's mother). It illustrates the fact that we are dealing with a throwback, a man who has certainly not evolved to the point that he is deserving of Belle. The Beast, on the other hand, is a far more promising companion, since he shows Belle that he can enjoy feeding the birds which live near his castle and he is able to take her snowball-to-the-face good-naturedly. Most important, he protects her from wolves at personal cost, dances with her and treats her with respect and love, matters of far more importance than the fact that he is physically misshapen and facing a possible future without love or a human face. He has not exhibited virtue in the past, but when he learns to do so he is rewarded permanently.

Irreducibly, the beauty and beast stories are about change: women who change by leaving home, men who change under the influence of love, women and men who change themselves by changing the way they see and deal with one another. Their prominence now illustrates the fact that we must deal with a great deal of change (particularly within our personal relationships) if we are to survive. The old stories of *normal* courtship and development have worn thin and are in short supply. These beauty and beast stories provide us with some assurance that we can not only endure the process of dislocating and destabilizing change but, indeed, thrive within it.

Also, in a society afflicted by interpersonal future shock, a society of anorexics, bulimics and products of the plastic surgeon's scalpel and liposuction tube, a society in which surface has often replaced substance, where material goods have all but effaced the essential, and where suitors pay more attention to the cut of their capes than the depth of their sincerity and love, the beauty and beast tales—like so many of our most important stories—urge us to look beyond appearances to the realities behind them. They urge us to think before labeling or categorizing, and to look beyond a person's appearance or previous history to the dimensions of the person's character and the contents of his or her heart. Virtues are far more important than conventional good looks or conventional wealth. Indeed, the individuals who are prized most may well be those who are either developmentally disabled or radically out of step with their societies.

Disney's Belle, for example, is considered quite odd by the other people in her village. She is bookish to a fault (in their eyes) and her lack of interest in Gaston (the darling of the village girls) is considered incomprehensible. Her love for her father, Maurice (an eccentric inventor) stands in sharp contrast to the scorn and ridicule which he receives from many of the townspeople, just as her love for the Beast seems strange to those whose first inclination is to attempt to kill him. Belle is anything but "normal," though she is, in every way, virtuous.

The same can be said of Clarice Starling. In the opening scene of the film she is followed by the camera as she runs the FBI obstacle course at Quantico. The camera convention suggests that she is being stalked as she runs through the woods. Only after several minutes of exposition do we *see* that she is an FBI cadet. The point is that she is doing something different; she is *someone* different. She stands out as one of the few women cadets and her short stature results in several tableaux in which she is surrounded by men who dwarf her.

She has lost her mother as well as her father and she has turned to a serial killer to help her find another serial killer. She has no time for a social life. When one of the entomologists at the Smithsonian (Noble Pilcher) hits on her she has to put him on hold until after she has completed her work on her case and completed her studies. The title of the story refers to her memory of the cries of the spring lambs being slaughtered on a ranch in Montana to which she is sent after her father's death. Again she is the outsider. She falls afoul of the authorities when she tries to save one of the lambs, striving unsuccessfully to rescue it from the knife. In the final confrontation with the serial killer she is alone. The FBI believe the killer is in Illinois; they break into "his" house with an entire team of agents. Meanwhile, Clarice is conducting her own investigation in Ohio and, unexpectedly, finds herself face to face with the killer. The contrast between their corporate action and her isolation (and, of course, inner resources) is reinforced by the director's cutting between the two scenes to both heighten the suspense and further delineate her character.

While she studies very hard and applies herself as best she can, she realizes that she is inexperienced. She also struggles doggedly against her sense of insecurity, something that Lecter picks up on instantly and is at first prepared to use against her. What saves her with Lecter is what saves her in the confrontation with the other serial killer: her character, her virtue.

Lecter's own situation is similar. When Clarice is asked the precise nature of Lecter's psychosis she responds, "they don't have a name for it." He is absolutely unique and that is his great burden. He holds God responsible for his nature (a subject to which he has devoted a great deal of thought, as his references to Marcus Aurelius reveal). In his bitterness he collects examples of church catastrophes such as instances in which ceilings have collapsed upon the faithful and he considers the marketing of a crucifixion watch with the second hand turning in Jesus' halo. He is not, however, the complete monster which his asylum keeper, Dr. Chilton, indicates, for Dr. Lecter, for all of his evil, retains certain basic virtues. He has been an avid student of human psychiatry, loves beautiful things (the Duomo in Florence, Glenn Gould playing the Goldberg Variations) and he is always polite.

He treats Clarice with respect, when Chilton, for example, does not. He may be a psychotic, cannibalistic, serial killer, but he is, simultaneously, a gentleman. In Clarice's world that counts for a great deal and it enables the two of them to work together with mutual respect. Lecter is not a pure hero, for his evil is too great, his virtue too little, but he is not an unadulterated monster. He shares with Clarice a succession of developmental disabilities and that—coupled with their mutual respect and mutual good manners—helps to establish a connection between them. She is a great heroine, in part, because of him.

Earlier we pointed out the social pressure that exists to institutionalize "normality." Positions have ranks, individuals have titles, churches have sacraments that are to be taken in sequence, fraternal organizations have degrees, and all of life can be broken into phases or stages through which we are expected to pass.

That, however, is not the complete story, for side by side with these institutionalized steps are a set of institutionalized forms of recognition which set the individual apart and highlight special accomplishment. The army, for example, has an entire set of ranks for enlisted personnel and warrant, commissioned and non-commissioned officers. At the same time the army has a set of medals which are awarded for uncommon service. Some are higher than others, of course, so there is some degree of ranking involved, but while every private is expected to become a private first class and then a corporal, no one is *expected* to receive the Congressional Medal of Honor.

This sort of award recognizes *difference*. It recognizes performance which is not ordinary but rather extraordinary. These kinds of awards do not recognize success within established, predicted norms. They recognize performance (as the military puts it) above and beyond the call of duty.

This kind of recognition can be found in all walks of life and in all areas of life. One may be a saint, for example, without having received certain sacraments. One can receive the Nobel prize without possessing a doctor of philosophy degree.

When one thinks about this it is striking how many such awards there are. Many of them come close to the pseudo-events of which Daniel Boorstin writes. Nevertheless, they are extremely important to us. For years, for example, the ABC evening news selected a "person of the week." Such individuals came from all corners of society. They were people who stood out, people who were exceptional. The segment must have been extremely popular. If it were not, the network would not have used it for as long as they did to hype the Friday news broadcast.

Every year *TIME* magazine selects a person of the year. This is always an occasion for controversy. Sometimes the person selected is evil or unappealing but yet consequential; that results in a flurry of letters from readers. Similarly, in those years when the magazine selects some *thing* rather than some person (the computer, for example), there is an instant controversy.

The Nobel prize awards are always major news, as are the Pulitzer prizes. Some institutions attempt to gain recognition for themselves by creating prizes which they then award. The winners (or losers) of some of these prizes (the Pulitzer or the National Book Award, for example) can be grist for great controversy.

Each year the Academy of Motion Picture Arts and Sciences awards show draws an international audience of staggering proportions, regardless of whether that year's crop of films is particularly impressive or not. A recent addition to the list of awards which receive special attention is the set of fellowship awards from the MacArthur Foundation. The press has quickly titled these the "genius awards" and those who receive them are immediately described as individuals who are decidedly different and special.

Within organizations we have teachers of the year and salespersons of the year. Once when we were buying a house our realtor informed us that his firm had an agent of the month award. The person selected was allowed to spin a dial which indicated a prize which he or she would then receive. Our agent showed us the dial and assured us with some pride that he had been asked to spin it often.

Employee of the month awards are important features in company newsletters and some organizations go so far as to provide the winners with additional perquisites, such as the use of a special parking place for the succeeding month. Everyone who endures might receive a watch upon retirement, but the parking space next to the front door is a more singular form of recognition and we see it as such. Company officials want us to see it as such, because they want to encourage performance that is extraordinary rather than ordinary.

Such awards cannot be equated with the recognition of heroic accomplishment, but they make a special point, an important point, namely, that while society encourages behaviors which satisfy expectations (and may discourage individuals who fail to meet standard expectations), it truly respects individuals who exceed expectations. These are often people who follow their own lights and listen to their own hearts. The prizes which they receive, both in self-esteem *and* in external recognition, are extremely important to all of us.

The same nuns who insisted that we follow certain forms of behavior *religiously*, also insisted that we *dare to be different*. Their point was that most would follow the path of least resistance or the path of maximum illicit pleasure and that only the brave, the heroic, and the "different" would follow their prescribed course. We now realize, however, that it sometimes takes great virtue to follow a different course than that which is generally prescribed, but that course can lead to heroism. Moreover, as we look at the institutions of society more closely, we realize that those institutions often encourage us to strike out on our own as well as to follow the more common paths. It is not impossible to do both.

There is a prejudice toward the "normal," of course. Otherwise we would not need the word "abnormal." We would use "extraordinary" in its place. The prejudice probably exists because it is often easier to be normal than to be extraordinary and those who are unlike everyone else carry a special burden of proof. They must convince us of the merits and the virtues of their way of proceeding and that is not always easy. When they succeed, however, we sometimes call them heroines or heroes. They need not convince *us*, however, to know in their own hearts what they already are and that, we should all realize, is what matters most.

Chapter V

Summation

> The characteristic of genuine heroism is its persistency. All men have wandering impulses, fits and starts of generosity. But when you have resolved to be great, abide by yourself, and do not try to reconcile yourself with the world. The heroic cannot be common, nor the common heroic.
>
> —Ralph Waldo Emerson

Refining the concept with a final, broad range of examples.

The study of heroes is as important as it is problematic. Heroes are not neutral beings whom we study like so many stained shapes on a microscope slide. They are important to us, in many cases desperately so. Heroes and heroines matter. They are the individuals after whom we hope to pattern our own lives, recognizing that we are likely to fall short of their accomplishments. In looking into their hearts, we realize, we are also looking into our own. Harold Lubin writes:

> If heroes are symbolic embodiments of our values and beliefs, then it is not surprising that a sober scrutiny of our own heroes can be a disturbing business—as is every investigation of the values and assumptions by which we live. The study of heroes is one way to learn about ourselves and our world by getting outside of ourselves and looking quizzically at what we are . . . . A thoughtful study of heroes will inevitably raise some of these most basic human questions: Who am I? How did I get this way? What do I want, and how shall I get what I want? What can I learn about myself and my needs from others, from my past, from man's past?
>
> (1968, pp. 4–5)

We often assume—certainly the attitude is made explicit in our culture—that success in life can be positively correlated with something called "normality" and with a process entitled "proper adjustment." Happiness, many would assume, is dependent upon it and those who are "abnormal" or "maladjusted" are frequently seen as struggling with psychological deficits which will impede their quests for happiness. Allowance is made for a certain degree of flair or panache among the demonstrably successful, but as a society we are given to voicing such maxims as "the bigger they are, the nicer they are" or that "for all her power and wealth and influence she is still a *real* person." The *real* person is *genuine*, which is to say, *like us*, which is to say, "normal."

We are often reminded of the mythological figure Antaeus, son of Poseidon and Gaia, who kept his power only when he remained in contact with his mother earth. Hercules lifted him into the air and crushed him. We praise those whom we consider "down to earth," individuals who have maintained their connectedness to their people, to their homes, to their roots. While nature exhibits her power from time to time in destructive ways, we are more comfortable with the cyclic and predictable dimension of nature's ways—the workings of what the songwriters Richard and Robert Sherman (in *Charlotte's Web*, 1973) call "Mother Earth and Father Time," the nature to which Ecclesiastes refers, implying that there are times and seasons for all things and a fundamental orderliness that underlies our experience.

The notion that success and happiness are somewhat dependent on a certain degree of "normal" experience, experience which is undergone in a predictable sequence, is not necessarily wrong. What interests us, however, is the fact that developmental theory can reinforce notions of normality, predictability, and sequential experience and subtly misrepresent the experience of those individuals whom society has often seen as unusually successful. The notion that "when you get right down to it he was just like everyone else" may not—in certain important ways—be true at all. When you get right down to it, many heroes and leaders may not be—in specific ways—like us at all. They may be deeply troubled, at times, perhaps, even deeply "disturbed," and their most dramatic accomplishments may not flow from their "normality," from their deep, genuine, common humanity, but from their manner of dealing with their maladjustment.

George Bernard Shaw (1930, p. 225) once commented that "the reasonable man adapts himself to the world: the unreasonable one persists in trying to adapt the world to himself. Therefore all progress depends on the unreasonable man." This is a predictable comment coming from Shaw, who was often considered an unreasonable man, but one of Shaw's close friends was T. E. Lawrence. Lawrence was, in fact, so close to the Shaws, particularly Shaw's wife, that when he was seeking anonymity he referred to himself, pseudonymously, as T. E. Shaw. Certainly many of Lawrence's activities in the Middle East can be characterized as attempts to try to adapt the world to himself. The same might be said of Churchill and Patton, both of whom—like Lawrence—were determined to bend the world to their own wills in order to save it. Jane Austen recreated her world in fiction in an effort to help us both see it more clearly and understand why that world required change.

The "unreasonable man" can, of course, be a madman as well as a hero. Hitler is commonly so characterized (and not without considerable justification), while we pride ourselves on setting up as his successful antagonist, Dwight Eisenhower: a pure, "normal" soldier from Kansas, a man who would not countenance swearing, a man whose military role was to bring together a disparate group and mold them into a unified fighting force—in short an administrator *par excellence*.

On the other hand, one could hardly characterize Patton and Montgomery as administrators and few would characterize them as "normal," though their contributions to the war effort were enormous. Leaders (such as Eisenhower) are not necessarily heroes and (as we will argue later) heroes are not necessarily leaders. In the film *Patton*, General Jodl, the head of the German general staff, comments contemptuously of Patton that now that he and Montgomery are fighting together the Germans will have to deal with "two prima donnas." Prima donnas are individuals who make strong personal demands, demands which come from their own enlarged egos and sense of their personal importance. Such individuals are, by definition, immature: pouty, self-absorbed, self-regarding. At the same time, they are selected for their role (that of the artistic principal) because of their great talents and gifts. They are, by definition, different. They are also the individuals whom the public pays to see: not the stage managers, not the administrators.

Developmental theory has certainly been applied to the lives of famous individuals and those individuals' problems, peculiarities, and private suffering have not gone unnoticed. Moreover, it has been argued that their triumphs have often resulted from difficult, even abnormal circumstances. However, as their lives are translated increasingly into myth we tend to block out whatever tarnishes our pristine image of them. The greater the distance from the so-called "great" the more difficult it is to see their feet of clay. Thus, it is sometimes believed that idealization serves the hero well; closer inspection jeopardizes his or her heroic status. This is related to the notion that no man or woman can be a hero to his or her valet, that no man or woman can be a prophet or prophetess to the members of his or her own village.

While on the one hand investigative reporters and authors of psychobiographies demonstrate the clay feet of our former heroes, the mythmakers take troubled individuals of heroic stature and reduce them to caricatures or cartoon figures with whom we can all be comfortable. It is a fact, for example, that Isaac Newton was almost completely humorless. It is said that he only smiled once in his life—when an individual asked him what was the use of geometry. Newton treated his academic appointment at Trinity College, Cambridge as a complete sinecure, sealing himself off from most human contact and never sustaining any personal relationship or even continuous correspondence with his colleagues at Trinity after he left there (Westfall, 1980).

This is not, however, the Newton we choose to remember—the diffident, reclusive individual who feared criticism and held off publication of his work, hoping that potentially negative reviewers might die first. Instead we opt for the genius-dreamer with laws of motion swirling in his brain, the demi-god mental voyager charting the mechanical dimensions of the cosmos for we lesser mortals. We see a scientist who gains his insight, his inspiration, the grist for his great synthesis, not from the inductive method of Francis Bacon, the cosmological work of William Gilbert, the experimental work of Galileo, the mathematical work of Isaac Barrow—all aided with the pivotal, self-developed tool of the calculus—but rather from the common, genuine, normal, human experience of sitting under a tree and having an apple fall upon his head. In our own time Einstein has been similarly caricatured, so that he might be brought down to "normal" size, most recently as a matchmaker bringing together young lovers in the film *I.Q.* (1994).

This is not to say that it is the developmental theorist who is to blame for turning the famous (and the infamous) into caricatures. Society does that very well without any professional help. (Pigeon holes may not be comfortable places for pigeons, as the literary historian Donald Greene has said, but they seem to give a certain amount of comfort to those who construct them.) It is true, however, that the attempt of the developmental theorist to chart a pattern of behavior that implies health, wholeness, and happiness undoubtedly contributes to the sense that there really is a "normal" way and to diverge from it is to risk the loss of something important.

This is not an inconsiderable point. We sometimes hear, for example, that "the great" are not really happy. They rise to great occasions. They perform heroic acts. They create magnificent art, etc., but they are not "truly" happy. Happiness, as Robinson Crusoe's father taught, is the particular benefit of those in the middle station. Such individuals suffer neither the wants and diseases of the poor nor the risks and pressures of the rich. This attitude is still common. While we may look up to the "great" momentarily, we quickly remind ourselves of the fact that they have often been denied the joys of genuine, which is to say, *normal*, human life. We speak of "poor little rich girls" and discuss "the loneliness at the top." While there is some truth behind these expressions, there is also a latent prejudice in favor of what we consider to be the *normal*, which is to say, the way "we" live.

Severe depression (Churchill's "black dog") finally overcame an elderly man whose brain could not, because of an impaired blood supply, continue to function efficiently. This depression, however, had haunted him for many years. In May of 1940, Anthony Storr argues, Churchill became the hero that he had always envisioned becoming. Many would agree that it was his finest hour. But in that dark time, what England needed was not a shrewd, equable, balanced leader. She was desperate for a prophet, a heroic visionary, a man who could dream dreams of victory when all appeared lost, a man who could merge his own desperate quest for happiness with his country's desperate quest for victory.

Churchill is a hero to a great number of people, but how many of those people would wish to undergo what he underwent in order to achieve what he achieved? As noted above, Yeats once said that if he had been happy he would not have written, he would have lived. We admire Yeats, but we realize that he never found the opportunity which history handed Churchill. Some agree with T. S. Eliot that Yeats was the greatest poet of the twentieth century, but few would trade their lives for his.

Or would they?

This notion of true happiness inhering in "normal" life has not always been blindly accepted. In a frequently quoted passage of *Walden*, Thoreau says that most people are in fact not happy. Most people, Thoreau tells us, actually lead lives of "quiet desperation." If Thoreau is correct, the mythmakers who stress the normality of the great are actually attempting to have it both ways—they are trying to acknowledge the special nature of the great while simultaneously denying it. "He is truly special, but when you think about it he is just like us."

Nietzsche argues that the slave makes a virtue of a necessity. He cannot act like a master and so he cherishes the virtues of the slave: patience, quietism, meekness, long-sufferance. Christianity, to Nietzsche, is thus a slave philosophy. It emerged because a people whose country was occupied by legions of armed Romans needed a philosophy which would see their condition as a happy one. Some of us continue that practice. In order to see our lives and experiences as normal and virtuous we construct a system of attitudes that correspond with the terms of those lives and experiences. Meanwhile, we lead lives of quiet desperation.

We would put the issue in the following terms. Society assumes that there is a "normal" structure of life (we are born legitimately; we outlive our parents; we marry; we have children who "replace" us; our children outlive us, and so on). Developmental theorists assume that there is a discernible *pattern* of life, progressing through stages and involving a set of key relationships, activities, and behaviors. The often unspoken assumption, then, is that this structure and pattern are somehow bound up with notions of health, wholeness, and happiness. There can be some divergence (indeed, everyone to some degree experiences that divergence) but it is still felt to be a very important thing to remain focused upon normative practices and patterns.

Those patterns to which theorists such as Erikson, Kohlberg, Piaget, and Freud might assent—as well as the "average" observer—constitute a course of definable experience which can be reinforced by science. Philosophers assert and observational scientists and survey researchers confirm. These patterns have also been represented in myth, in fiction, and in popular culture. This results in a cultural consensus that asserts that certain behavioral and experiential patterns are both healthful and desirable.

However, when we look at the experience of the highly successful, we often find correlations between that success and what we have termed developmental disabilities: breaks in the "normal" pattern, gaps in the individual's experience, flaws in the individual's personality. Freud, after all, also argued that imaginative activity springs from dissatisfaction, not satisfaction. The problem, of course, is how is one to think about this phenomenon? How is one to deal with a correlation between pain or loss or absence and success? Are we, for example, to inflict such pain in the interest of encouraging such accomplishment? Should we condemn our children to loneliness and isolation, knowing that that can sometimes become the impetus for creativity?

Three years after his father's death Churchill wrote, "Solitary trees, if they grow at all, grow strong; and a boy deprived of a father's care often develops, if he escapes the perils of youth, an independence and vigour of thought which may restore in after life the heavy loss of early days" (Manchester, 1983, p. 188). In his biography of John Churchill, the first Duke of Marlborough, his famous ancestor, Winston wrote that:

> Famous men are usually the product of an unhappy childhood. The stern compression of circumstances, the twinges of adversity, the spur of slights and taunts in early years, are needed to evoke that ruthless fixity of purpose and tenacious mother wit without which great actions are seldom accomplished.
> (Manchester, 1983, pp. 188–89)

Thus, Churchill saw a portion of his own experience in that of the man who had defeated the French in the War of the Spanish Succession, two centuries earlier. Winston himself had been born in the great palace which a grateful queen had given her Captain General for his success in defeating their bitter enemy. But if the loss of a parent or the neglectful behavior of a parent can lead to triumph it can also lead to pain, suffering, and, in some cases, failure. There are no guarantees. Even if it can be shown that childhood suffering *sometimes* leads to great achievement, we would hardly consider provoking such suffering rather than preventing it in the hopes of stimulating such achievements.

The institutionalization of abnormality would be, to say the least, bizarre if not cruel. There are instances of such attempts to which we might point. For example, it is universally acknowledged that while personally somewhat unattractive, Douglas MacArthur was unquestionably the finest cadet in his West Point class, gaining grudging acceptance of the fact from even his deepest enemies. One day, within a very brief period of time, MacArthur was somehow able to transport one of the cannons at Trophy Point, a prominence on the north side of the post, jutting toward the Hudson River, to an "impossible" position on one of the stone buildings hundreds of yards away. The action has never been explained scientifically and as late as the 1960's an attempt to solve the riddle of MacArthur's method remained a class problem in the cadets' engineering curriculum. Since the Army was originally first and foremost a corps of engineers, MacArthur's prank (comparable to the notorious pranks perpetrated by the students at Caltech) can be seen as a kind of military feat. If, for example, one of Longstreet's divisions could have expeditiously gotten its cannon on the slopes of Big Round Top on the second day of the battle, there might never have been a third day at Gettysburg.

The point here is that the Academy drew from this experience of MacArthur's the conclusion that truly great leaders possess (as part of their standard psychic equipment) a flair for the unusual, a taste for the dramatic, exceptional act. Why? Because Doug MacArthur possessed such a flair and he ended up wearing five stars. Thus, while the Academy enforces standards of conformity in curriculum, dress, demeanor, attitude, and thought it is also unusually tolerant of the occasional divergence from that conformity.

The focus on "normal" progression may be important for purposes of description and analysis, but it carries with it an obvious risk. To the extent that we are aware of our own "abnormalities" we are tempted to fixate upon them and excuse our personal shortcomings because of them. Looking constantly toward the past and reminding ourselves of pain or deprivation we forget to move forward and instead risk cultivating a personality of blame and complaint. Patton might have done this, but he chose instead to look to the future. He was always determined to move forward, both in his personal life and in his military tactics. In the Schaffner film he announces that he will relieve the encircled 101st Airborne at Bastogne, not fall back and regroup, as he is advised. "I don't like to pay for the same real estate twice," he responds.

The belief, in short, is that "normal greatness" includes a certain amount of abnormality. In the Ronald Maxwell film *Gettysburg* (1993), Longstreet comments that southerners like their generals to be both religious and a little mad. Hence, even within an extremely regimented and disciplined system, the United States Military Academy has been not only tolerant of a certain degree of abnormality, but has even encouraged it, since abnormality is thought to be a normal contributor to greatness.

Two divergent attitudes are somehow made commensurate with one another. On the one hand, a profession is created with the most rigid of career paths: officers proceed through set stages, "getting their tickets punched." The careers are managed by branch chiefs who are actually termed Career Management Officers. At the same time, it is assumed that the path to greatness is not always straight and narrow. "Ticket puncher" is a term of opprobrium, comparable to "careerist." It is not a compliment. Having perceived zigzag patterns in the careers of the greatest generals, the Army has adjusted its notion of normality to encompass the unique and the exceptional, as, for example, in their tolerance of special articles of dress for general officers.

The corollary is that developmental theory must do the same. While structure and predictability are good and in some ways necessary, most of us experience zigzag patterns which—as in the case of the hero—may occasion triumph as well as pain. Our notion of the normal must encompass a certain degree of the abnormal and while heroes and heroines might sometimes undergo exceptional pain, their exceptional accomplishments can serve as sources of inspiration to the rest of us.

They already do so serve, of course. The one thing that can be asserted with certainty is that we need and require heroes. "History," Marshall Fishwick writes, "is meaningless without heroes; there is no score before they come to bat" (1968, p. 1). Those heroes, we would hasten to add, might be unsung as well as well-known. Dramatic historic moments are often associated with the actions of the prominent, but historic movements depend upon the actions of the unsung as well. Sometimes the unsung later become prominent. Because of Michael Shaara's book *The Killer Angels* we are now familiar with the actions of Joshua "Lawrence" Chamberlain, who held Little Round Top, securing the Union left flank on the second day of the battle, but how many can name the engineer who secured Culp's Hill, protecting the Union right?

The identification of heroes is not without risk, of course. Sidney Hook, for example, realizes that false gods and false prophets will tend to arise with some regularity and he warns us that we must be careful in choosing those whom we are determined to follow. Charisma is attractive, but it is also potentially dangerous, as history reminds us time and again.

Nevertheless, it now often appears that we have swung too far in Hook's direction. History and its heroes are now systematically demythologized and we are left with a world that is gray and impoverished. "A land that has no more legends," it has been said, "is condemned to die of cold." This is not to say that we should believe lies or emulate individuals who were actually frauds, but it does suggest that we require models for our own behavior, even if (perhaps especially if) those models' experiences transcend our own both in their nature and in their intensity.

Today, many of our youth have lost the sense of heroism which can both direct our lives and give coherence to our history and its traditions. Bruce Wilshire comments that "Few students have any sense that they are missing empowerment and stabilization which comes from identifying with revered persons in the past; as if there were a constant voice which said, these are our sources, our very substance, and we can count on them!" (1990, p. 12) The result, Wilshire observes, is a profound lack of trust among the young. They have not negotiated Erikson's first stage. They are detached from their history and, as a result, detached from their society and even from themselves. Heroes—themselves the frequent product of some form of alienation—could serve to ameliorate the alienation which those who could observe them now feel.

The quiet desperation of the many might be tempered by the realization that those who have achieved greatness have often done so, not just because of some trans-human effort or quality, but because of the depth of the pain from which their actions have sprung. There is, in short, the realization that their wounds have not only healed but proven to be the source of much of their success. This can lead to the belief that our own wounds might actually become the healing forces in our own lives, just as they have been the healing forces in the lives of our heroes and heroines.

We hear a great deal about the personal failings of those who were once considered great. Given the proclivities of the popular press, these failings are often of the most banal variety, falling under the broad umbrellas of greed and lust. The result is that some conclude that there are no heroes, that "heroism" is a construction of idolaters and mythographers, and that, once we truly discover the dimensions of an individual's experience, we will find a failed and fallen person who is, at bottom, just like the rest of us.

We are not discounting the importance of the so-called "character issue" here. Some of our putative heroes have indeed been the products of the publicist's art and it is appropriate that we express our disappointment when the actual nature of their character and personality is revealed. The character issue is, in fact, pivotal. As we argued earlier, the hero may fail the stages; he or she *cannot* fail the virtues. There is a difference between disability and an absence of character. Heroes are often those individuals whose virtues enable them to triumph over their disabilities, whose disabilities are sometimes the springboards for their accomplishments.

Thus, rather than despair of the possibility of finding heroes and heroines, we must broaden our notion of weakness or disability, balance it with the requirement of virtue, and look for those elements in the lives of the heroic—those personal narratives or segments of narratives—which can inspire us through example. It is precisely our own pain, our own perceived disability, which both make the heroic example necessary and—through our common link with the hero—make it possible. The hero may be exceptional in his accomplishments; he is generally not exceptional in his possession of disabilities. Our need for models, our hunger for inspiration, is, in part, a desire to see evidence that pain and disability can be transcended. The hero not only transcends suffering; he demonstrates the fact that not only can suffering be survived, it can become the source and occasion of triumph.

A number of recent biographies offer examples of the kinds of experience to which we might point. Each is instructive in a different way but each underlines the basic point.

Alan Turing—the subject of Andrew Hodges' moving biography, *Alan Turing: the Enigma* (1983)—was one of the most remarkable people of the twentieth century. A person who, like all of us, was in need of family love and attention, Turing was often separated from his parents because they lived outside of England as tax exiles. His earliest and in some ways his deepest friend, Christopher Morcom, died prematurely of bovine tuberculosis at the age of nineteen. Turing was openly homosexual and, after being subjected to what can only be characterized as governmental persecution (persecution that required him to ingest certain drugs or face imprisonment, the drugs resulting in his growing breasts) he committed suicide in 1954.

Alan Turing was also a Fellow of the Royal Society, the father of the modern computer, and the individual who led the successful effort at Bletchley Park to break the German code during World War II. Turing was always considered different, and not just in his sexual orientation. He was generally disheveled in his appearance. His teeth were stained. He was a loner, one capable of indulging in bizarre actions, pedaling his bicycle, for example, over uncommon distances. He inhabited mathematical realms which few others could enter. However, in a time of national crisis, he focused his unique attributes and made a contribution to the war effort the importance of which proved to be incalculable. He serves as an inspiration to all of us who have felt different, alone, and apart, but he is a special source of inspiration to those who seek to demonstrate that patriotism and a particular sexual orientation are not incompatible.

In his book, *Certain Trumpets* (1994), Garry Wills discusses the question of leadership and invokes both examples and counter-examples to make his points, the principal one of which is that one cannot lead without followers, that leaders and followers exist in a symbiotic relationship in which each requires the other for self-definition. Wills argues that Ludwig Wittgenstein—the subject of a recent biography by Ray Monk (1990)—cannot be considered a leader because of the fact that he had no followers. He may have been one of the two or three greatest philosophers of the twentieth century, but he was still not, Wills tells us, a leader.

Wittgenstein is another individual who, like Turing, underwent a succession of excruciating experiences. While the family was Jewish they concealed their heritage and Ludwig was baptized Catholic. All of his brothers but one committed suicide. That brother—a concert pianist—lost an arm in the First World War. In fact, Ravel's *Concerto for the Left Hand* was composed for Wittgenstein's brother, Paul. Despite this loss, his brother continued to play. His family's response to his playing (with one arm) was that it lacked taste and was too extravagant. The Wittgenstein family was not one of exceptional warmth and understanding.

Wittgenstein, who did not actually speak until he was four years old, was also homosexual and he also contributed to the war effort on behalf of his country, serving in the Austrian army during World War I. (He was, in fact, recommended for the Austrian equivalent of the Victoria Cross and he received his country's second highest medal.) Wills is correct that in his philosophy Wittgenstein's practices discouraged would-be followers. He was notoriously slow to publish, when he did publish. Much of his work consists of notes rather than finished discourses. His philosophy is also extremely complex and recondite; it is not subject to simple formulation or interpretation. Moreover, one of his cardinal concerns is the limitations of philosophy. Hence, when he saw that philosophy was heading toward a dead end he resigned his academic position at Cambridge and took up other work.

In the Vienna telephone directory he referred to himself as an architect. Indeed, he was a skilled architect and engineer. During one phase of his life he worked as an elementary school teacher, living in a remote village where no one had any idea that the local schoolmaster—Herr Wittgenstein—was, in fact, one of the most important philosophers of the century. Clearly he was capable of achieving Eriksonian generativity through this experience.

Wittgenstein *did* lack followers in Wills's sense of the term. He was not a leader. He was a hero. He did not lead people; he inspired them. He inspired by the awesome strength of his intellect and he inspired because of the strength of his character. He believed that the only way that we can really improve the world is to improve ourselves and he pursued that task with nearly monastic zeal. The fact that he followed his beliefs and left his academic position when he came to doubt the usefulness of the philosophic enterprise is a source of inspiration to those who share his doubts but lack his courage. The fact that he would seek out other, useful work, work that was beyond the limelight, is similarly inspiring, like his earlier decision to give up his portion of his family's wealth and pursue the life of the mind unfettered by material distractions. The fact that there are so few men like him made it difficult for him to be a leader; on the other hand, his very uniqueness—his triumph in the face of parental coldness, war, the denial of his religious identity, and the violent self-destruction of his siblings—makes him, to many, a hero.

Earlier we referred to General James Longstreet, Lee's "old war-horse," his most trusted general. Longstreet is the subject of a recent biography by Jeffrey D. Wert (1993). In Longstreet, once again, we have a man of great accomplishments and uncommon disabilities. He was practically last in his class at West Point. (George Pickett—confederate general and leader of 'Pickett's Charge' at Gettysburg—*was* last in his.) Longstreet also suffered the deaths of three of his young children (to scarlet fever) early in his life. Of his ten children only five survived to adulthood. Throughout his life he exhibited both great competence and great insecurity with regard to his reputation.

It is Longstreet, of course, who has been forced to carry much of the brunt of the criticism of the confederate strategy at Gettysburg, though the strategy (recently reassessed—Carhart [2005]) was Lee's. Once it was clear that the federal army was on the high ground along Cemetery Ridge, Longstreet advised a movement around the federal left, positioning the confederate army between the federals and any number of major northern cities: Philadelphia, Baltimore, or Washington. The federal army would be forced to pursue, lest the confederates capture one of these prizes and exacerbate the growing anti-war feeling in the north. When the federals pursued them the confederates could choose their own ground for battle and achieve another victory comparable to that at Fredericksburg, where they—not the federals—commanded the high ground and the stone wall.

History, of course, has proven Longstreet to be correct, though it must be said in Lee's favor that the six-day absence of J.E.B. Stuart's cavalry made it impossible for the confederates to know the precise location and size of the union army. Had they moved to their right and headed east they might have run into tens of thousands of troops. We know now that they would not have done so, but Lee did not know then what we know now.

Nevertheless, Longstreet has, in some ways, been scapegoated for Gettysburg, the most important battle in the American Civil War and one of the most important in the history of warfare. The more accurate term for "Pickett's Charge" is "Longstreet's Assault," though it was ordered by Lee and Longstreet was so pained by the impending slaughter that he could only nod when Pickett asked him for the order to attack.

Later in his life Longstreet made a crucial error of judgment, running for president of the United States, in the south, as a Republican. (He had also converted to Roman Catholicism.) Thus, the official response to Longstreet at the end of his career was one of deep bitterness, the southern loyalty to Robert E. Lee unending. Upon the occasion of the great reunion of the Army of Northern Virginia, Longstreet was shunned. He was not even invited. When Longstreet appeared anyway, in full uniform, he was greeted warmly by his troops.

His personal judgment was sometimes flawed and he often struggled to rewrite history, remembering battles in such a way as to present his own performance in the most favorable of terms, but at bottom he was a superb soldier of uncommon bravery, whose sense of strategy and tactics was far in advance of that of his contemporaries. His personal weaknesses, combined with the fact that history and circumstance did not cooperate in such a fashion as to highlight his abilities, serve to make him an interesting hero. His disabilities were not strange or unusual; they were common and all too human. On the other hand, his accomplishments were so great that he enjoyed the abiding love and respect of those who served under his command, despite the fact that their cause was ultimately lost.

Longstreet's (and the south's loss) contrasts with Lincoln's victory, but that victory was hard-fought and hard-won and the personal price paid very high. Lincoln's private suffering was intense and ongoing and ended only with his death at an assassin's hand. The loss of his children, the pain endured by his wife, the challenges of his presidency and his personal, lifelong depression cast dark shadows over his experience. That suffering may, however, have contributed materially to his success. Joshua Wolf Shenk subtitles his recent book, *Lincoln's Melancholy* (2005), '*How Depression Challenged a President and Fueled His Greatness*'.

Melancholy, or depression, was viewed differently in the nineteenth century than it is today. Following the humor theory of antiquity and the renaissance, melancholics were viewed as specific human types with black bile dominating their physiology and generating their personality. However, their melancholy was seen more positively than depression is today in that it was associated with the capacity for intense self-reflection. Regardless of the viability of humor theory, it is clear, Shenk argues, that Lincoln was a person of deep self-reflection, who constantly pondered himself, his world and his role in it. His desire to affect that world in positive ways was one of the principal motivators in his life, the personal call to duty and action serving as a counterbalance to his melancholia. In the nineteenth century melancholia was associated with a realistic (if sad) view of human realities; Shenk writes of Lincoln's 'depressive realism', a vision of experience that was certainly appropriate to his time of bloody civil war. Before one can solve a problem one must understand it and confront it; there is no doubt that Lincoln's awareness of the nature of his and the nation's situation was profound.

As with Churchill, Lincoln's personal struggle against private adversity prepared him for the struggles against the greater, public forms of adversity which he faced, as president. Like Churchill, he compensated for his melancholia with a deep sense of humor. Like Churchill, he took refuge in poetry and his literary self-education and the resulting facility with language made him one of the great speakers and communicators in human history. Like many depressed individuals, he sublimated his feelings and found relief in creativity. His depression also fostered a sense of humility which is one of his hallmark characteristics. First and foremost, it forced him to focus intently on larger goals and helped him to save the union.

A man closer to our own time who functioned both as a leader and as a hero is enjoying a resurgence of attention now. Sir Ernest Shackleton (1874–1922), one of the notable heroes of the twentieth century as well as one of the great leaders of men, had reason to feel that his work was not fully appreciated by his contemporaries. A member of the 1901–1904 Antarctic expedition of Robert Falcon Scott, Shackleton suffered the ignominy of being sent home (for medical reasons) prior to the expedition's completion. An Irishman and a member of the Merchant Marine rather than the British Navy, Shackleton *was* often given short shrift by historians in their rush to remember and praise Scott.

In recent years this has changed, as a large number of books, exhibits, and films have been released, celebrating Shackleton's accomplishments. Shackleton has finally become a household name, though his greatest adventure ended in failure and near disaster.

In his expedition of 1907–1909, the *Nimrod* Expedition, he ascended Mt. Erebus, located the south magnetic pole and reached a new Farthest South latitude of 88° 23' S, a point only 97 miles from the South Pole. Shackleton's South Pole party also discovered the Beardmore Glacier route (named after Shackleton's patron) to the South Polar Plateau, and were the first persons to set foot on the plateau. Later, in 1914, Shackleton and his crew sailed to Antarctica in their famous ship, the *Endurance*, to begin the Imperial Trans-Antarctic Expedition (with the approval of the First Lord of the Admiralty, Winston Churchill, who had declined Shackleton's offer to put the *Endurance* at the country's disposal for the war effort). Their hope was to enter the Weddell Sea and cross the polar region to the Ross Sea, where they would meet their supporting ship the *Aurora*. In Shackleton's judgment, the crossing of the South Polar continent from sea to sea remained the great object of Antarctic adventuring, since Amundsen had already conquered the South Pole. This journey of nearly 2,000 miles was not completed. Despite their best efforts, the unforgiving ice closed around the *Endurance* in 1915, shattering her hull and leaving her crew cast adrift on a vast ice floe. In 1916 Shackleton was able to bring his men to safety at Elephant Island, a voyage of approximately 180 miles. From there, he and a small number of his crew (five men in addition to Shackleton) sailed across the Arctic seas some 800 miles to South Georgia Island. The waters between South America and Antarctica known as the Drake Passage that Shackleton had to cross are among the most treacherous in the world. The skies were so gray and overcast that Frank Worsley, Captain of the *Endurance,* was only able to take celestial navigation readings at four points during the 800-mile journey.

Upon his arrival at South Georgia Island, Shackleton and his party were forced to cross the rugged interior of the island on foot, reaching a whaling station on the island's north coast. Had they not made this trek they would have surely perished. Shackleton then proceeded to return to Elephant Island to rescue his crew members left there, none of whom lost their lives.

While Scott is often touted as a brilliant and courageous explorer, it must be remembered that he died in his attempt to reach the pole, while Shackleton was able to rescue each and every member of his crew. In fact, in his journey to South Georgia Island, Shackleton took with him two difficult members of the crew (the carpenter, McNeish and the bosun, Vincent) whose presence might have jeopardized the morale of those left behind on Elephant Island.

The enormity of Shackleton's accomplishment can only be appreciated by seeing the *James Caird*, the boat in which Shackleton and five others sailed the 800 miles to South Georgia. The *James Caird*—named for one of the supporters of the expedition—was just over twenty-three feet, a tiny piece of wood amid the swells of the icy Arctic seas.

Shackleton's personal life was often uneasy, his marriage nearly stretched to the breaking point (by his long absences, his philandering, his questionable investment schemes) and his principal goals were never achieved. What he did achieve and exhibit, however, was a strength of character and an ability to command the loyalty of his men that were truly exceptional.

Upon their return to Britain many members of his crew were immediately sent to fight in the First World War, where some perished, a poignant end to the story (then), but now the exploits of Shackleton and his crew are celebrated around the world and the man who sometimes lived in Scott's shadow has now achieved the acclaim he so richly deserves.

Such examples could be multiplied endlessly. One might, for example, add a particularly instructive narrative, that of Eleanor Roosevelt, for many one of the great women of the twentieth century. Even those who disagree with her political orientation can admire her energy, her dedication, and her tenacity.

Her challenges were many. They are among the principal foci of Doris Kearns Goodwin's 1994 account of the Roosevelts' marriage and its implications for the life of the nation during a time of war.

She was, of course, uncommonly homely, a feeling shared by and communicated to her by her mother. Her father (like her brother Hall) was a severe alcoholic who, among other things, had made a servant girl pregnant. He died at the age of thirty-four after suffering a seizure which followed an attempted suicide. To protect her own feelings, her mother retreated into a cold aloofness. Eleanor married not only an only child but an only child with an indomitably domineering mother, one who lived with them in the White House and openly vied with her daughter-in-law for her son's love and attention. Franklin, of course, was unfaithful to Eleanor, having had an affair with Lucy Mercer, a woman in Eleanor's employ. Later in his life Franklin was extremely close to his secretary, Marguerite "Missy" LeHand, though opinions differ as to whether or not their relationship was also a physical one. Eleanor's own feeling with regard to sexuality was that it was an ordeal to be borne rather than a pleasure to be enjoyed. In fact, as Garry Wills notes, her feelings were so strong in this regard that she forced her daughter to endure metal cage-like devices on her hands to prevent her from touching herself.

One source of emotional support to her was the friendship of Harry Hopkins, but when Franklin utilized his abilities in the administration of the lend-lease program, Eleanor felt as if her old friend had deserted her and her social-justice agenda for the prosecution of the war, a necessary evil.

How did she survive this situation? Essentially, she took the occasion of Franklin's relationship with Lucy and his later relationship with Missy as opportunities. Distancing herself from Franklin, she focused all of her interest and attention on questions of social justice. Similarly, her realization that Missy was prepared to look after every detail of Franklin's personal and professional life freed her from the obligation to attend to the head-of-the-household 'wifely' duties from which she also sought relief.

Franklin's political career could not have survived a divorce and his mother had threatened him with disinheritance if he ever attempted to seek a divorce; hence the power relationships within the family shifted as a result of his infidelity. Like Jane Eyre and the blinded Rochester, Eleanor and Franklin became equals with his weakened position and she was able to redefine the nature of their partnership.

There was still great warmth between them, for Franklin continued to find Eleanor intellectually engaging and he saw—in her political activities—the kinder, gentler, more idealistic side of his own political orientation. She was, for him, of great political importance, since she helped secure votes for him which increased the size and diversified the makeup of the coalition upon whose support he depended.

This is not the sort of marriage which all would choose (though one sometimes hears references to the Clinton *partnership* as well). However, it is a clear instance of an individual with great strengths who has survived great pain in order to attempt great actions for the public good. Those experiences which might have crushed other individuals were seized as opportunities and put to fruitful use. (Nelson Mandela was less lucky, his former wife becoming a significant political liability rather than a political and personal asset. Another true hero, he survived this succession of events just as he survived extended imprisonment and other painful experiences.)

Had the pattern of Eleanor's experience been more "normal," her responses to her husband's actions might have been of a different order altogether. Instead of being destroyed by her experiences she attempted to use those experiences to chart a course which was commensurate with her own personality and intellect as well as with her own goals. She achieved what she achieved because of her character and because of the fact that her disabilities could be turned to useful purposes. If she was uncomfortable with Franklin's physical attentions, she was yet capable of directing and defining her life in such a way as to free herself to do what she most wanted to do—not to do or to attempt to do what would conventionally be termed normal.

The recent loss of Princess Diana underscores the fact that individuals who carry profound psychological wounds can nevertheless take on heroic stature. Indeed, as her brother pointed out in his moving eulogy, Diana's own feeling of unworthiness enabled her to understand the feelings of others, particularly those who lived in deep pain at the peripheries of fame and the public consciousness.

She was the most frequently photographed woman in the world, but she was perfectly positioned to understand the plight of those most likely to be overlooked or forgotten by that world. Capitalizing on her own feelings and depth of understanding she was able to relieve the pain of others. The public response to her death was so broad and deep that it is clear that she had become a heroine and a model for many.

Her personal life consisted of a series of psychological traumas, from her parents' divorce and neglect to the discovery of her husband's infidelity, her own divorce and her struggle to defend herself in the face of the power of the palace and press. Betrayed by those who should have loved and cared for her, she turned to others in need and did what she could to ameliorate their condition.

While her own suffering and death occasioned global mourning, the pattern of her experience also brought great hope to a significant portion of suffering humanity. Seeing the developmental wreckage that was her personal life we can also see the extent to which her experience was a determining factor in both her philanthropic work and in her relentlessly conscientious efforts as a parent. Her personal wounds became the animating force behind her personal and public efforts and her virtues (particularly her courage, her tenacity, and her supreme sense of empathy) were always in evidence.

What engages us in each of these life stories is the human narrative: the personal journey whose lessons can both seize our imagination and strengthen our spirit. The relationship with the inspirations of fiction and myth is clear. We see *patterns*, patterns of experience that inspire us, patterns that are not "normal," patterns which consist of triumphs in the face of pain, suffering, and *ab*normality. Early in her life Eleanor Roosevelt was forbidden to read in her bed before breakfast. Defying her grandmother's rule, she hid books beneath her mattress, waking up at 5:00 a.m. to read such writers as Dickens and Scott, Dickens who defended the poor and sought their relief, Scott who recreated a lost world of romance and chivalry in which the good sometimes triumphed over the evil.

We too find inspiration in these narratives—whether they be fictional or historical—as we seek to find, in the lives of our heroes and heroines, evidence that we too can survive our psychic wounds and perhaps even utilize them to achieve our personal wholeness and experience our personal success.

Ever since the renaissance we have been imbued with the notion that our civilization is progressing in an upward cycle. We continue to progress because, it is argued, we stand on the shoulders of giants, enjoying the fruits of the accumulated knowledge and wisdom from antiquity forward. The idea of progress, as it has been termed, was stimulated by the early successes of sixteenth- and seventeenth-century science: Copernicus' astronomical studies, William Gilbert's studies of magnetism, Harvey's discovery of the circulation of the blood, Galileo's studies of falling bodies, and Newton's physico-mathematical description of the universe.

In the eighteenth and particularly in the nineteenth centuries the concept of evolution was joined with the idea of progress. Social Darwinists argued that societies and civilizations evolve, just as species do. The discovery of Cro-Magnon man and the expanding anthropological and archaeological record suggested a comparable pattern, one that reinforced the notion that the development of the species was best thought of as a process of ascent, with early hominids developing the use of hand tools and our near ancestor Neanderthal man being succeeded by homo sapiens sapiens.

In psychology one can see a similar pattern. Beginning with Freud and continuing with the work of such developmentalists as Piaget, Kohlberg, and Erikson, there has been an effort to theorize the quest for human happiness and wholeness as a progressive movement entailing the successful negotiation of a sequence of developmental stages.

When one examines the lives of heroes and heroines, however, one finds something quite different. These individuals whose lives represent exemplary patterns to others, have not always successfully passed through the predicted (or idealized) developmental stages. On the contrary, their lives are sometimes so fractured psychologically that one might speak of their having suffered from developmental disabilities—failures to negotiate particular developmental stages.

However, the fact that they have suffered in this way and yet triumphed as men and women, triumphed to the degree that they are now seen as models, suggests that our theories of personal developmental progress require adjustment. Indeed, in many instances, the developmental disabilities of the hero or heroine have been the occasion or cause of that individual's heroism. Their wounds, one might say, have served to heal them as individuals. Their wounds are both the cause of their ultimate success and the source of our admiration for them.

In short, those individuals whom we often take as our heroes and models have led lives which do not follow predicted or idealized developmental patterns. The extent to which they have achieved greatness in spite of their psychological disabilities is not only the cause of our admiration; it is also an occasion for us to examine our own developmental disabilities and consider the manner in which the experience of the hero might offer us the basis for hope.

To the extent that each of us feels that his or her life has not followed an ideal pattern, we can take heart in the experience of the hero and see the possibility that our psychological wounds might do more than heal. They might also be the cause of a more generalized healing. Surviving our disabilities could, in effect, lead to a different (superficially "abnormal") pattern of experience which, far from associating us with the permanently flawed, actually allies us with the heroic. Thus, the hero's triumph can be the source of our own inspiration. He or she is not just an individual who does things with exceptional courage or skill; he or she does these things in spite of or as a result of ostensible weakness or disability.

The pattern of experience exemplified by the hero is, then, not a predictable story, but a story constructed by the hero when the normal life plot is somehow fractured. The hero does not proceed unimpeded along a smooth and straight road. The hero's journey is frustrated or disrupted, but he or she goes on to construct a new life narrative, to pave *his* or *her* road, one different from the common path.

In studying the developmental process Erikson wisely availed himself of a multiplicity of materials (literary, biographical, and historical, among them). He was a man familiar with and comfortable with the arts and he shrewdly adopted a broad view of that which was potentially useful to him. He did not, for example, exclude the use of survey research or other highly empirical materials, just as he was prepared to learn from the experience of single, special individuals such as Luther or Gandhi.

Erikson studied and adduced *patterns*. One of our purposes in the preceding pages has been to look at the developmental pattern charted by his research and attempt to add some nuance to it based on the special case of heroes and the manner in which the experience of the hero can be particularly instructive and inspirational for those of us who have experienced developmental disability without seeing the full implications of those occurrences and patterns.

It is the *pattern* which, finally, seizes our attention and grips our imagination. Aristotle noted in his *Poetics* that drama was an imitation of human action. It does not pretend to mirror or replicate human action. On the contrary, it creates a separate action which can help us to understand the actions of real life. The classic "Aristotelian" plot is one that has long sustained the writers of drama and fiction. Essentially, that plot concerns the experience of an individual whom we find interesting and good but also flawed and human. In tragedy the flaws may be of a deep and pivotal nature. In the course of the story the protagonist is subjected to conflict which intensifies as the narrative develops. Eventually that conflict reaches a climax which leads to a resolution which is either happy or, at least, moving and instructive.

Conflict is essential. The "normal" is not enough. Two conclusions can be drawn. The first is that we need stimulation which requires exaggeration. These stories then, take us away from the normal ruts which constitute our lives and pique our interest as they stretch our imaginations. Thus, narratives help us "escape" through their unreality. A more nuanced conclusion would be that these stories imitate the actions of our own lives but on a different imaginative level. Art helps us to understand our selves and our lives by highlighting and intensifying those thoughts and acts which make up everyday life.

Our own everyday lives are marked by conflict and we realize our own human flaws and vulnerabilities. We hope that our personal narratives can reach a happy conclusion, but we have no illusions about the neatness of the process. We know that there will be setbacks, missed opportunities, and gaps in our experience.

Sports and games imitate life; they concentrate the experience of life, as Huizinga long ago argued. If we look at board games which represent life in linear fashion (*Snakes and Ladders*, *Monopoly,* for example) they are always replete with the possibility of setbacks. We move from point *a* to point *b* as expeditiously as possible, depending to a considerable degree on luck, but we prepare for the false step, the wrong-colored space, which diverts us from our goal and returns us to a previous position. The setbacks are not a rare or uncommon occurrence; they are part and parcel of the game.

Heroes appear in fiction as well as in life. One of the secondary meanings of the term is that of the protagonist (the "hero") in a fictional narrative. As art concentrates life, highlighting and intensifying experience, the pattern of the hero's life (in fiction and in history) inspires us by exemplifying virtue, often in the face of setbacks and challenges.

The world of romance, for example, is one in which actions are often not completed. Spenser's *Faerie Queene* (1590–96), taking its impetus from the generalized ethos of renaissance epic, makes the point that virtuous knights do all that they can in a fallen world, but they never succeed in fully conquering evil. Dragons and necromancers remain, making mischief and causing pain in the face of the heroes' and heroines' efforts to eradicate them. There is no tidiness here, only the ongoing attempt to champion the good in a world where the good is continually threatened.

In our own time the impact of romance is felt in crime fiction and, occasionally, in science fiction. Final triumphs can never be expected. The most that we can hope for is virtuous action and we realize that such action is often achieved at great personal cost and in a cause that might finally prove to be lost. Victory cannot always be attained, but virtue can always be practiced.

We would suggest that the need for heroes can be seen in the art that we choose as well as in the lives that we study. That art, however, is changing. For many today the artistic patterns which dominate culture are musical. The young as well as the old are enclosed in the portable space delimited by earphones, the verses of modern poets and rap musicians echoing in their consciousness and shaping their vision of the world and their approach to experience.

Printed verse has often been displaced by melodic verse and the thoughts, hopes, dreams, and aspirations of modern artists fill our airways and (sometimes) expand our spirits. These works give voice and expression to the thoughts of their generational listeners and help to embody those listeners' feelings and beliefs. The songs often speak to issues which bear on heroism, just as the singers of the songs have often taken on heroic trappings in the eyes of their audiences.

Athletes—as important now as they were in Aristotle's time—exhibit skills and are often idolized as a result, but the idealized future they offer their worshipful followers is more a future of fame and material success than a future of glory or spiritual fulfillment. The Olympic spirit is an ideal; the commercial realm is, more often, the reality. Musicians, on the other hand—those other special heroes of a post-print electronic age—speak to both head and heart. Their words and melodies—often taken to refer to their personal experience—reside in our brains as wired memories, returning whenever we need them. Their art repeats and repeats; we possess it both spiritually and neurologically.

Given the nature of some contemporary art and music this is not a situation which we can look upon without some measure of concern. The messages of anger and loss are not always balanced by messages of hope and joy. If we look to the heroic life, the balance is achieved for the hero—who may suffer anger and loss among many other pains—finally rising above them and healing himself and/or others in his or her journey. The hero stands before us as model and pattern, offering us an outstretched hand of hope. Sometimes the singers of songs do the same, but not always, and some have been struck down by their own hands as well as by circumstance. The impact of such events indicates how important the singers and songwriters are; it also indicates the extent to which such an impact is generation-specific. Once again, the personal timeline is of crucial importance.

While not bounded by cinematic or print material, the representations of the hero (in film, myth, history, or biography) lead us into the future by offering us considerations of the past. This may be why we sometimes hear people say that the age of heroes is past. It is *not*, of course, but the hero is part of the world of legend as well as the world of fact, part of our cultural legacy and common inheritance.

In the post-print culture of the early twenty-first century we run the risk of losing part of this legacy. Thus it is important to remind ourselves of the crucial and vital role that heroes can play in meeting human needs and in lifting the human spirit.

It is interesting to note that the four historical figures whose experience we have particularly examined shared a number of things which we are in danger of losing: a sense of personal destiny rooted in the past but reaching into the future; an intense awareness of the importance of education (particularly when it did not come easily to them); a deep knowledge of human history and the manner in which it both anchors our experience and enriches it; a perception of the importance of landscape, both natural and human, and a realization of the fact that landscape helps to define us as a race, as a people, and as individuals; finally, a driving desire to achieve an identity which embodies strengths and virtues and is characterized by an abiding integrity.

If one examines the developmental stages through which these individuals passed we see an interesting set of commonalities. Each of these individuals (except for Jane Austen, a special case) had difficulties with four of Erikson's stages (the second, fourth, fifth, and eighth stages) and both Lawrence and Churchill suffered through stage one. On the other hand, all three of them displayed great initiative and great generativity (stage three and stage seven). As we stated previously, one can fail the stages but not the accompanying virtues. One can be psychologically orphaned, as Churchill was. One can be one of the greatest voices in human history with regard to the relationships between men and women, as Jane Austen was, but die a spinster. One can lead armies, but be unable to achieve personal intimacy, like Lawrence, or lead armies while playing a studied role, like Patton, a role that was constantly threatened by self-doubt concerning his ultimate worth. Each of these notable leaders triumphed because each of them possessed the virtues, each possessed a particularly strong sense of purpose and direction and a desire to care for those who followed them. The virtues are crucial, the developmental disabilities often causal, the impact of the two in tandem essential.

In effect, these four individuals (Lawrence, Churchill, Patton, and Austen) looked to the past and to the future: to the past for inspiration and direction, to the future for hope and an enduring tradition. It is our privilege and our task to heed well their example in this regard, to search out our own pasts, to dream of our own futures, and to move through our present with strength and virtue. Their experience and example are part of a vast web of associations, images, words, and attitudes which serve to inform us, to inspire us, and to help illuminate our vision.

Chapter VI

Becoming a Hero

> Nurture your mind with great thoughts; to believe in the heroic makes heroes.
>
> —Benjamin Disraeli

How can I become a hero or heroine to myself and to others? Am I already one?

In the preceding chapters we have discussed the fact that many of our heroes appear to be individuals who carry developmental wounds, wounds that have, at times, not only healed but been a source of strength and identity for the individual in question. Recently we have witnessed a parallel phenomenon in American life: the description of individuals in terms of the specific form of oppression under which they have labored.

This is related to the issue of "identity politics," the notion that your particular experience of the world determines your political posture within that world. This is related to the notion of what has been termed "victimology" insofar as individuals have often grouped themselves under individual banners which specify the particular forms of suffering which they have endured and which separate them from the more privileged majority. The discrimination and violence which these individuals have suffered is, thus, a pivotal part of their experience, a determiner of their identity, a shaping force in the development of their political posture.

This set of attitudes heightens the claims of subjectivism which are part of the legacy of post-renaissance epistemology and strengthens the claims of relativism which are a special legacy of the nineteenth century. Since the victims of the majority have often been culturally marginalized, it is argued that they should be especially vocal in the articulation and celebration of the particular views which they embody and espouse and that the study of the individual cultural products which flow from their experience should receive particular attention, displacing the cultural products of the majority which, it is argued, embody the attitudes and beliefs which have contributed to the suffering of minorities.

For some, this phenomenon has signaled an increase in cultural richness. For others it has signaled the arrival of a "culture of complaint" which has proved to be divisive rather than diversifying. Whichever position one takes, it is clear that these attitudes have been central to the so-called culture wars of the 1980's. They continue today in a number of fora, from the structuring of public policy to the structuring of the humanitics curriculum in the contemporary university.

From the point of view of human development a number of observations might be offered. For example, before we focus too much attention on the experience of subsets of humanity we should remind ourselves that there remain a large number of basic human experiences which cut across cultures. Kant was at pains to demonstrate the existence of a specifically *human* way of understanding. To be human is to perceive in a certain way; it is as if (some have said) we are all born with a particular form of spectacles that we are powerless to remove. We share common physiological traits, common abilities and limitations *vis à vis* other creatures, and common needs—for sleep, for water, for clothing, for food and for shelter. Ultimately, we share a common gene pool. We do not see as fish and flies do. We do not hear or smell as dogs do. As infants we are absolutely dependent on our parents or caregivers. We do not stand and walk moments after we are born. We utilize tools and technologies to an uncommon degree. These facts are true across cultures, countries and times. As many have pointed out, there is really only one human *race*. If we believe Chomsky, that race is hard-wired to acquire the use of language, one with a common grammar.

Moreover, the sufferings of individual subsets of humanity can be conceived in more general terms. While minorities suffer from specific forms of discrimination and prejudice, for example, the configuration of the possible sets of "minorities" changes with context. Everyone, more or less, has suffered from some form of discrimination and prejudice. African-Americans have been prejudged, but so have women with blonde hair. Gays and lesbians have been prejudged, but so have celibates and Mormon polygamists. Southerners have been prejudged by northerners, midwesterners by easterners, Californians by everyone. Middle easterners are stereotyped in action-adventure movies, but so are most fathers in situation comedies. The uneducated are patronized by the educated, the short are teased by the tall, and everyone is corrected or chastened by Parisian waiters.

The point is that one might focus on prejudice, discrimination, and stereotyping without plunging into subjectivism or relativism. These are unfortunate human experiences which all of us have undergone (though the intensity of the prejudice and the intensity of its impact, of course, varies widely). While the degree of suffering may certainly have been different, this form of injustice is common and the stress of its commonality may be more useful in mounting opposition to it than the stress of division and difference.

This is not to deny the existence or discount the importance of cultural differences. What you see is often a function of where you sit. The Roman empire looked very different to patricians than it did to slaves. Christopher Columbus looks very different to Italian-Americans than to Native Americans. Moreover, there are very real differences between people based on long-term cultural experience. Sociobiologists, for example, would point to millions of years of hunter-gatherer divisions of labor and note the resultant differences between women and men. Women, for example, see better in the dark and hear across a broader range of the sound spectrum, while men have a better sense of direction in broad land navigation.

Nevertheless, there are irreducible human traits and experiences which we share in common. Subjectivism and relativism must be kept within bounds. If there were not some semblance of a common humanity and common human experience all communication (and, hence, all education and all civil society) would be impossible. Thus, while subjectivism and relativism are enjoying a striking degree of currency today, we must remind ourselves that their denial of an absolute "human nature" does not necessarily also mean the absence of a common humanity.

Everyone has had the experience of a great feeling of commonality with some individual who would be pigeon-holed in a very different slot from one's own presumptive slot by the supporters of a politics based on identity. Gay and lesbian individuals, for example, may find much common ground with straights and, thus, bond with them, just as some African-Americans, Asian-Americans, Native Americans and other members of the human family will have strong points of commonality with some whites and, hence, bond with them. This suggests several things. First, we are far more complex as individuals than the schematic outlines of identity politics would suggest. We might be gay but also conservative. We might be black but also fundamentalist Christians. Our views of the environment might be more important to us than our views of religion and vice-versa. Individuals may see their profession as more determinative of their identity than their gender or ethnicity. Clearly, many straights prefer the cultural products of gays, just as many whites prefer the cultural products of blacks. Russell Jacoby (1994) has pointed out that for all the protestations of "difference" we are far more homogenized as a culture than we were a century ago. At the turn of the twentieth century the proportion of immigrants was far higher as a percentage of the American population than it is today. There were far more foreign-language dailies then than now. Intermarriage was far less common then than now. In the culture of the twenty-first century, for example, it is quite common to encounter individuals who are the offspring of "interracial" marriages, with half-brothers and half-sisters who are the offspring of a still different combination of parents. We regularly meet Asian-Americans with such names as Kevin or Jennifer who are dating Hispanics and who prefer Italian food and black music. It is curious that identity politics has had such a strong impact at a time such as this when a blurring of such differences (with such notable and prominent examples as Barack Obama, Tiger Woods, and Derek Jeter) is often the order of the day.

Many distinctions of class and culture are extremely relative and alliances of country, state, or neighborhood can weigh heavily in the articulation of our sense of identity. These can shift *vis à vis* the context in which we are defining ourselves. To one group we are "old" but to another "young." To some we are "short" but to others "tall." To some we are seen as "rich" while others may see us as "poor." To some we are "Texans," to others "Americans," to others residents of Ft. Worth rather than Dallas, Presbyterians rather than Episcopalians, etc. All of these can shift based on context, location, time, and the person or persons with whom we are being compared. One well-known basketball player used to say that since he was not blessed with height he had to make up for its lack with skill and speed. He was 6'8". We once knew a rabid New Yorker who had grown up in Queens. He said of another person of our acquaintance, "He's not a real New Yorker; he's from Albany."

It is also clear that cultural commonalities of various kinds can predominate over the subset commonalities of identity politics. For example, it is sometimes assumed that certain subsets of the human community will be defined by specific aspects of their difference. However, this is simply not the case when we examine our own experience. We find instead that individuals who have been labeled "different" from us are individuals with whom we might share a great deal. The "difference" may thus be trivial or, to put it another way, the commonality might be far more important. For example, members of the military (black and white) may feel far greater commonality with one another than with civilians of the same race. Born-again gay Christians may feel more commonality with born-again straight Christians than with gay atheists.

While this perhaps seems obvious, it is important to reiterate the dangers of narrowly characterizing our fellow human beings. We believe it is also dangerous to define ourselves in terms of our wounds or differences when that self-definition might lead to a fixation on the wounds rather than a focus on the potential for growth within our unique experiential context. Self-pity is not a firm basis for human development, just as the continual stress of difference erodes the possibility for the further building of a community of shared humanity.

Rather than seeing ourselves as wounded individuals deserving of pity we should see ourselves as individuals capable of not only transcending our pains and hurts but capable of building upon them as well. Those elements of our experience have served, in part, to shape our experience, and the outline of our experience is, in part, the basis for our identity. This is who we are. However, our wounds should be seen as tempering us, not serving to diminish us perpetually.

The question is, how does one do that? How does one follow the example of the hero and capitalize on that which has made us unique? Earlier we discussed the characteristics of the hero and the needs which the hero satisfies. We spoke of the manner in which our sense of the heroic is dependent on our historical perspective point as well as the perspective point of our personal timeline. We discussed the fact that heroes exist in different spheres of experience and that while there are national and even historical heroes there are also countless heroes whose lives are intertwined with our own even though they might not be familiar to people beyond their narrow sphere.

In particular we talked about virtue and the fact that while heroes may "fail" some steps in the developmental process they do not "fail" the virtues which are associated with the stages in that process. The next logical question follows: can virtue be taught?

This is a question that has challenged many thinkers, a question to which there is no simple or easy response. Looking back over our own education we recall the Catholic articulation of the issue. There *are* (we were taught) four cardinal virtues (*cardinal* because pivotal): prudence, justice, fortitude, and temperance. We are taught them by Wisdom, a gift of the Holy Spirit (who brings other gifts as well—understanding, counsel, fortitude, knowledge, piety, and fear of the Lord). There are moral virtues such as chastity and truthfulness and theological virtues (faith, hope, and charity). The latter are shared with divinity; human virtues are rooted in them and they both inform and give life to the moral virtues.

Virtues are acquired by education, they are acquired through deliberate acts, and they are acquired through perseverance. Education is simple enough. Virtues can be enumerated, they can be defined, and examples can be given which clarify their nature and illustrate their practice. Deliberate acts are also simple enough. We understand a virtue and we ourselves practice it. Perseverance, however, is not always so easy. The catechism tells us that it comes through divine grace. Virtue is defined as "a habitual and firm disposition to be good," but "dispositions" do not come easily. And what of divine grace? How can that be acquired? (The catechism also stresses that the moral virtues are purified and elevated by divine grace.)

Leaving aside the issue of divine grace and the manner in which it might be given or received, we can secularize the message of the catechism and argue that, in the development of a *disposition* to be good, humans (of whatever faith) need help. Virtue results from habitual behavior (as Aristotle argued), but it is not easy and it is not automatic. If it results from action repeated over time, can it be taught? It is one thing to describe it, quite another to inculcate it.

Plato's response is in some ways inconclusive, but it is nonetheless suggestive. In the *Meno* Socrates argues that since there are no teachers of virtue one must conclude that it cannot be taught. In part this is a commentary on contemporary teachers; in part it is a reflection of Plato's views on learning and knowledge. For Plato knowledge is essentially reminiscence (because of our prior existence) and cognition is a form of recognition. Socrates says, explicitly, that "there is no teaching, but only recollection."

For Plato the soul is "virtuous" when it acts according to its own, complex nature, as Fulton Anderson (Plato, 1983) argues. The soul could be seen as a powerful system in which the constituent elements need to be kept in a state of equilibrium. That equilibrium results, in part, from virtue. Appetite is balanced by the virtue of temperance, passion or temper by the virtue of courage, and reason is balanced by the virtue of wisdom. Habit, Plato argues, serves as the basis for moral action. In effect, we become what we are by repeatedly doing what we do. His statement in the *Meno* is interesting: "Virtue is neither natural nor acquired, but an instinct given by God to the virtuous." Aristotle's formulation is similar:

> Now some think that we are made good by nature, others by habituation, others by teaching. Nature's part evidently does not depend on us, but as a result of some divine causes is present in those who are truly fortunate; while argument and teaching, we may suspect, are not powerful with all men, but the soul of the student must first have been cultivated by means of habits for noble joy and noble hatred, like earth which is to nourish the seed.
>
> (1941, p. 1109)

Once again, we need help. There is a famous Graham Chapman (of the Monty Python troupe) comedy skit that consists of a "class" on how to play the clarinet. The audience is promised a complete lesson on the instrument. Once having "attended" this class the students will totally understand what one must do to play the instrument. The instructor steps forward with a clarinet in his hands and says, "This is how you play the clarinet—you blow through this end and move your fingers over these holes and levers."

Virtue is somewhat similar. One can have a sense of that of which virtue consists, but not then be able to immediately practice it. The practice of virtue requires habitual activity as well as knowledge and models. Moreover, it requires patience and tenacity. It may well require the help of others and, at certain levels, divine grace or its secular counterparts.

We do not pretend to have a solution to a question which Plato could not resolve, but we would suggest that some tentative responses are possible. Virtues *can* be described; examples of practitioners of virtue can be adduced; the knowledge and, hence, practice of virtue is neither easy nor automatic but it can certainly occur over time. It may require external help and support. At times this support can be direct and straightforward. At other times there is something so special about that help that we might describe the experience in more spiritual terms. Without alluding to questions of faith or to questions, for example, with regard to the creative process, we believe that one can speak of the possibility of an experience which some would term *inspiration*. Certainly we use the term every day and the nature and degree of the process of inspiration can be realized across a spectrum of meaning, from the quick click in the brain that reminds us of something momentarily forgotten to the deep, nearly inexplicable experience which comes unannounced and changes our lives forever.

We would argue that heroes might play some role in this process. Heroes instruct us by providing patterns of experience and models of behavior. They clarify the nature of virtue by embodying it in practice. They face challenges and surmount obstacles in such a way that their actions can be seen on a generalized human level (as if they were characters in a great allegorical play) as well as on a specific, individual level (as if they were personal friends or acquaintances). They are individuals whose behavior we seek to emulate. At the same time, our deeply personal heroes move us in special ways. Their experience resonates with our own; their challenges remind us of our own challenges; their triumphs suggest that we might also, in some measure, share in their success. We see something of ourselves in their lives, some hope for ourselves in their achievements.

In short, they inspire us. They provide us with spiritual sustenance, with the moral and psychological wherewithal to attempt to model our behavior on theirs. We ask ourselves a series of questions: What would they think of us? How would they assess or measure our actions? If we were to meet them could we hold up our heads? They are, in a secular dimension, our saints, our local or household deities, perhaps even our angels.

While this may seem to be a psychological stretch, particularly when the hero is especially austere or remote, it is far easier to see when the hero is closer at hand. If an individual tells us that her greatest heroine has always been Joan of Arc we might raise an eyebrow, but we would not do so if she informed us that her greatest heroine has always been her grandmother. Toward the end of the *Nicomachean Ethics* Aristotle discusses heroic action and the fact that its possibility is not restricted to the powerful and prominent. He writes that "we can do noble acts without ruling earth and sea; for even with moderate advantages one can act virtuously (this is manifest enough; for private persons are thought to do worthy acts no less than despots—indeed even more) . . . . "

We discussed earlier the point that developmental psychology has tended to establish a pattern of predicted experience involving the negotiation of steps or stages. Success on that model consists of an orderly, "normal" progression through those stages. Success is then equated with happiness or at least a sense of completeness and satisfaction. At the same time, we noted that many of our heroes and heroines have missed some of those steps or failed to progress beyond one or more of them.

The experience of the hero or heroine then gives us heart, for we ourselves feel, to a greater or lesser extent, that our experience has not been entirely "normal." The hero's success, however, suggests that we need not lose hope. We might achieve heroism if not normality. At any rate, normality may be less interesting or attractive to us than we first thought, once we study the experience of the hero.

There are other points that one might make with regard to this issue. Developmental psychology emerged at a time when normality was cherished as a social value. It came at a time when change was more gradual and, hence, more easily absorbed. A sense of order and balance in the individual was easier to describe, since it was, to a considerable extent, mirrored in the external society or, in some cases, coerced by it. What Alvin Toffler termed "future shock" had not yet made its full force felt.

That has now changed and we find ourselves in a world where the constant flux once described by Heraclitus is one of the dominant aspects of human life. We now experience centuries' worth of change in a matter of weeks and months. Moreover, the writers of this book and a large number of its readers are individuals whose personal lives have bridged worlds of cataclysmic change. We remember a very different past because of our direct experience of it while, at the same time, we live in the present and anticipate the future.

Our children and grandchildren have not experienced a world without television. We have known such a world as well as the world of 500-channel cable systems. Our sense of duration and our sense of the speed of change remains a phenomenon that separates us from our forebears. To take a single example, the industrial revolution is often dated from 1760± to 1820±, a period of sixty years. The communications revolution, on the other hand, proceeds at a far faster pace. We all remember our first experience with the computer, with the fax machine, with the iPod and the cellular phone and we have "absorbed" them with far greater ease or far less hooplah than attended the absorption of the automobile or the airplane.

We have grown accustomed to having our feet in two (or three) different worlds. While we live in a world of chaos theory we are not terribly distant from a world of Einsteinian relativity or Newtonian mechanics. While we now take highly imagistic electronic technology for granted, we continue to use print technology in forms that would be immediately recognizable to a London bookseller of the late 1480's or a newspaperman of the 1690's. We remember the social stability and the sense of authority once represented by established churches but we now regularly encounter religious polarization and the destabilization represented by single-issue voting involving specific beliefs. While we might fear resurgent fundamentalisms we also see resurgent forms of worship, belief, and behavior previously associated with the pagan, the primitive, or the superstitious. New Age bookshops market music, crystals, and angelology books and these flourish in a time of skepticism and even cynicism. The "New Age" section of the local chain bookstore may well be larger than the "Philosophy" section.

We still see and still experience long-term committed relationships at a time when divorce is extremely common; we remain parts of families or extended families when "friends" or "networks" have sometimes replaced the nuclear family. We value family-owned businesses and corner stores even though we live in an economy of multinational corporations conducting global business. Millions use the interstate highway system, but many still remember the transportation system which preceded it. Comparatively inexpensive and widely-linked jet travel is now commonplace, though many remember pulling themselves up to their seat row on the DC-3. Multiple methods of travel may be employed in a single trip and multiple methods and technologies jostle in our memories, their individual advantages being reconsidered as our contemporary experience shifts.

We value nature even as we watch its displacement by cyberspace. We grow and give flowers even as we study fractal geometry or manipulate digitalized images. We speak of our "home town" even though we conjure with a pattern of constant travel and relocation that has now come to include global relocation. We still speak of our "careers" but we have become used to the fact that three or more separate careers may now have become the norm.

In short, we continue to grow and change, to carve out a life and structure a pattern of experience, in spite of the fact that we have less of a sense of "family," of "home," of national borders, or of the constraints of time or space than were once common and commonly accepted. Life expectancy norms are far more fluid; gender roles are vague, courtship rituals are uncertain and our sense of stability in relationships, in beliefs, in organizations, in work, and in institutions has eroded. *Normal* is not a word which carries the meaning which it once did, particularly for those who remember key aspects of the world, society, and economy which preceded those with which we now must deal.

It is useful to remind ourselves of that fact, to remind ourselves of our resilience, to remind ourselves that we have achieved great moments of joy and happiness and success in an ethos that has been anything but "normal." It is useful to remind ourselves that Freud's century was also Darwin's and that adaptability is a survival technique that should not be underestimated. It is a technique which nearly all of us, willy-nilly, possess.

At the same time that we reassure ourselves of our ability to negotiate unanticipated obstacles and fashion our personal lives within a society with fewer (or very different) reinforcing structures, we should remind ourselves of the tools at our disposal for recentering ourselves and achieving a sense of stability and continuity reserved in the past for a select few.

One of the difficulties of the modern world is the constant possibility of a disconnection between appearance and reality. While this is a theme which has haunted poets and novelists for centuries, its importance has grown in an era of richly imagistic communication systems, political doublespeak, and what Daniel Boorstin has termed pseudo-events ("ladies and gentlemen, welcome to the *first*, *annual*, vegetarian chili pageant"). We have also lost some of our anchoring (or had it redefined) because of the growth of international business and the expansion of the operations of transnational corporations. A shopping mall in Santiago, Chile is now largely indistinguishable from a shopping mall in Santa Rosa, California. The golden arches of McDonald's have been installed across the globe and Colonel Harland Sanders' smiling face looks down upon us from signs in nearly every world capital. Video game screens consist of iconography which translates across cultures. We seldom know the city or country in which the game is set. All that we know is that we are in a dangerous section of town and people are fighting with one another. In video games as well as in television, in books and in films, the world of science fiction, the world of gothic horror, and the world of swords and sorcery have now intersected with our "own" world and changed it forever.

Nevertheless, there are irreducibly real aspects of our experience which are now more accessible than ever before. After the battle of Gettysburg was fought and won, Lawrence Chamberlain commented on the fact that people would later come to southern Pennsylvania to visit the site of the battle. He said that the battlefield would be a "vision place for souls," a place where one might see, hear, and feel things which would alter one forever. For all of the kitsch (from the chicken and dumplings at General Lee's Headquarters [recently closed, mercifully] to the offerings at General Pickett's Buffet [still flourishing]) Chamberlain has been proven correct. The scene is continually visited, studied, and reconceptualized. The new Museum and Visitor Center, a $100M+ facility, depicts the causes and consequences of the Civil War with authentic artifacts, archival materials and 21st century technology. The battlefield and the cyclorama painting of the battle are undergoing constant refurbishment and restoration and the number of hotels and motels within the area has proliferated dramatically. At night it is common to find people sitting on Little Round Top, listening and watching, just as every morning and afternoon one can see a line of visitors who walk off the long mile between Seminary Ridge and Cemetery Ridge, this time unimpeded by cannon, carbine and musket fire.

The earth, the trees, the rocks, the graves, and the shrapnel remain and they are all available to be seen. We sometimes forget how easy travel has become. In the eighteenth century travel was a luxury of the rich and it largely remained so until very recently. In the nineteenth century "travel writing" often bordered on science fiction in the extent of its exaggerations and the number of its tall tales. Very few could challenge the claims made by its authors and many of its readers often read their work for its fantasy value. The first units of the large motel chains still dot our landscape, side by side with the small motor courts that preceded them. Even with recent increases, air travel is remarkably inexpensive when compared with the days when only the very wealthy could afford to travel. We can visit Gettysburg, we can visit Culloden, we can visit Waterloo or Gallipoli; we can visit the invasion beaches at Normandy and experience our history with all of our senses. Indeed, with technological enhancement, we can experience it even more deeply. The site of the Battle of Hastings, for example, is now equipped with digital devices which not only explain every aspect of the battle but also present those aspects from multiple points of view. You can listen to the Norman perspective or the Saxon, or, if you choose, you can experience the battle from the perspective of a woman of the eleventh century. Similarly, you can travel to Bayeux and see the tapestry which illustrates the battle (a fitting, intermediate point on your way to the World War II invasion beaches).

In the eighteenth century English students completed their educations by following their studies with a "grand tour" of Europe. In later generations students took a senior trip and then went to college, spending their junior year abroad, but that practice is now becoming increasingly dated as more and more students regularly travel and study abroad. The novelty is gone since the experience is so common. Our work-study students who attend college with the help of loans sometimes work overtime to save the money required for their airfare to Europe for spring break. With the competition at major hubs in below-peak periods it is often less expensive to fly to Europe than to airports in America less than one-fifth the distance away. An Irish student who studied with us recently had a few weeks open between the completion of her master's degree and her return to Europe. She promptly packed a knapsack and toured the Andes.

The past is an integral part of us, not only in the history which has shaped us but in the gene pool which we carry. We can visit our more distant pasts by exploring such sites as the caves and rock shelters of the Dordogne and the Valley of the Vézère. Lascaux is now closed to the general public, but a precise replica has been built a few yards from the original cave. Countless other caves remain open, including that at Pech-Merle on the river Lot. If Lascaux is the Sistine Chapel of prehistory, Pech-Merle is its Chartres. Standing in any of these caves, in silence and in faint light, one can see drawings that are tens of thousands of years old, drawings that stood on the same site tens of thousands of years before Christ, tens of thousands of years before Aristotle, tens of thousands of years before the Trojan war.

We read about cave bears in the writings of Jean Auel, but we can see actual cave bear pits in the Rouffignac cave. In a number of caves we can see negative handprints from our prehistoric ancestors, some of them swollen with arthritic joints. At Pech-Merle there are what are claimed to be prehistoric footprints. When several of the caves were originally discovered, animal bones (the remains of the meals of the artists) were found beneath the drawings. The remains of scaffolding have also been found.

Thus, while we read the accounts of our ancestors by modern evolutionary psychologists we can connect with them in more direct ways. In Les Eyzies de Tayac, the center of prehistory, one can receive instruction in the making of primitive tools and stone materials are available to help reify the process. A few yards down the street one can visit the site where Cro-Magnon man was first discovered.

While a shrinking world alters our sense of reality it also deepens it. We may not doubt that historical events occurred or historical personages lived, but the ability to visit sites and see artifacts directly materially changes our relationship to those realities. In some cases our skepticism is sharpened, in others our imaginations are expanded. Sites may be modest compared with our expectations concerning them (Hissarlik/Troy, for example, though new evidence suggests that the site was far larger than earlier digs have suggested). In many homes of prominent individuals the presence of the occupant may still be strongly felt (Freud's house, for example, though he lived there for a brief period of time, Samuel Johnson's, Darwin's, Churchill's and Jane Austen's).

Having seen with our own eyes, Johnson would say, our knowledge is more certain and more authentic and with the growth of our knowledge comes the growth of our self-confidence. We are no longer at the mercy of the mediating commentators whose voices we were previously forced to trust. Seeing things directly enables us to see and feel other things. Standing by the Aubrey holes which form an outer circle beyond the main stones at Stonehenge we are connected with the seventeenth century and the man who discovered them. We suddenly discover Stonehenge through Aubrey's eyes as well as through our own. Turning and seeing the burial mounds of the chieftains along Salisbury Plain or approaching the monument from a distance we see Stonehenge in a different context altogether, a context that cannot be captured in book illustrations or postcards.

Sutton Hoo, the seventh-century burial ground outside Ipswich, is now open to the public. Adjoining the mounds covering the burial sites of the kings and worthies of East Anglia are the burial sites of executed criminals, condemned to sleep in unhallowed, pagan ground by later secular/religious authority. The entire sweep of history from the dark ages to the present lies beneath the visitor's feet there.

Why should this be important to us? Quite simply, it is important because we are a part of history and history is a part of us. Our sense of ourselves, particularly our sense of our capabilities and possibilities, is shaped by our reading of the past and of the experience of the men and women from earlier times whom we have come to see as models. Human development is a process; the practice of virtue is a habitual process; the cultivation of the individual requires a vision of experience as well as a set of discrete experiences.

That vision is deepened and expanded whenever our knowledge and experience are expanded. It is deepened and expanded whenever our understanding and appreciation of other human models grows. It is deepened and expanded when we are physically and psychologically positioned in a wider sphere. It is said that we like open architectural space because we expand, psychologically, to fill it. Thus, we walk through a great railway terminal like that in Washington, Philadelphia, Cincinnati, or New York, listen to the click of our heels and its echoes from marble walls and we feel more real, more authentic. We feel larger. We feel present. In Florentine palaces we feel larger still.

Size alone conveys a heightened sense of reality, but so too does the insistent presence of history. Churchill's command bunker, now a part of the Imperial War Museum, has narrow hallways and low ceilings, but the presence of real telephones, real code books, real beds, real reinforced cement ceilings, and especially real maps conveys a sense of desperate effort, desperate danger, and the possibility of survival and ultimate triumph.

Put simply, we learn more, we feel more, we understand more, and we become more. The extent and cost of modern air travel makes this available to more people than at any time in human history and adds a dimension to the possibility of personal development that would be unimaginable to our ancestors. It is one thing to study prior human experience in an effort to improve ourselves, quite another to become a physical part of that experience virtually at will.

It should come as no surprise that developmental theory would focus upon steps and stages. Such patterns dominate the structures of our lives and of our society. Associations confer recognition by offering individuals "membership," while others are "associate members," still others "founding members." States are given "most favored nation" status or excluded from organizations until they mend their ways.

Within families privileges are tied to age. Hand-me-down clothes are the legacy of the later-born, while eternal "babyhood" is sometimes conferred upon the last-born. It shocks us when we see brothers and sisters in their middle years when all have not aged equally gracefully and the age demarcations are less clear. (They remain clear to the brothers and sisters.)

Athletic teams have captains and co-captains, fraternal organizations various levels of status or commitment. Within the church are deacons, priests, monsignors, auxiliary bishops, bishops, archbishops, cardinals, and popes; within the military there are ranks from private to general (and general ranks from brigadier to general of the army). Universities have instructors, lecturers, adjunct professors, assistant professors, associate professors, professors, and titled professors.

There is no question that, to some degree, we must progress through stages. We have to in order to survive. Human children are born with an exceptional degree of dependency. The mother is omnipotent, the newborn child the center of the universe. Many of our later psychological challenges are traceable to this reality. Children must leave home if they are to become adults; children must replace their parents as their parents replaced their grandparents.

Deceased leaders must be replaced. Societies must strike out in new directions under the leadership of new generations. The need for mature leadership and the simultaneous need for innovative, youthful leadership guarantee that there will be a certain degree of stress and friction within all human systems. Upon these basic realities a great deal of human literature and mythography has been built. The books which we consider masterpieces are often the books which have successfully embodied accounts of our quests as individuals and as members of a society. It is interesting to note that most of our most popular forms of genre fiction (science fiction, crime and mystery fiction, romances) go back to pre-novelistic forms (pastoral, chivalric romance, comedy) which articulate a conception of reality which is essentially mythic and grounded in developmental issues.

Thus, our conception of life is affected by our conception of art, for each reflects the other. In our art, as in our vision of human development, we are imbued with a deep sense of sequenced drama. Much of our recent literature, however, has added nuance to this vision. For example, the ethos of renaissance epic (Spenser, for example) includes a vision of an incomplete world, a world in which evil is only temporarily suppressed, a world in which knights must learn to sacrifice as well as to exhibit courage, a world in which heroism may involve far more behaviors than a simple progression through stages ending in success and satisfaction.

American writers emphasized this point by creating hardboiled detective fiction, a form that looks to that world of romance, where English mysteries ("cozies") look to the world of comedy, a world in which evil is expelled and all is eventually set right again. In the American version the world is only temporarily set right and the protagonists are often lonely and *incomplete*: often without pasts, without wives or husbands, without families, without settled work, steady income, or honorific titles.

While others have enjoyed and appreciated this vision, there are elements of it which are distinctly American. America, it has been noted, is the only major industrial country in which citizens are not obliged to carry identification cards. The crime and suspense writer Robert Ferrigno said that for a time he "lived without paper," that is, without identification, without government documents, without any obligations beyond the need to meet the cost of daily living.

In short, while our society is replete with steps and stages, titles and designations, resumés that depict "progressively more responsible experience" and uniforms which change with our office, there is, built into our society, the notion that it need not be that way. There are different opportunities—opportunities to strike out in unconventional directions, opportunities to try on new identities, to attempt different lifestyles, to slow down or to speed up, to drop in or to drop out. Most of us will experience both of these aspects of life simultaneously. We require both stability and change, peace and challenge. The heroes whose lives we have examined have not been individuals whose lives are totally separate from the patterns imagined by developmental psychologists. However, they are individuals whose lives have not been bound by such patterns, individuals whose pain and challenges have become opportunities rather than impediments, the basis for accomplishment rather than for self-pity.

How can each of us emulate these individuals or the individuals we would select as our personal heroes and heroines? How can we serve as heroes and heroines to others? How can *we* inspire *them*? In the first place we should remember that we are always being observed. How do we wish to be seen? As an object or as a person? As a target or as a model?

As we suggested above, each of us lives two lives, a personal life which may not square with the patterns and structures which others expect us to follow and a more public life in which we are measured against such patterns and structures. The personal life may be a life of great struggle and pain, but it can also be a life which is informed by a combination of virtues which can inspire others. The public, "normal" life may be a life of conventional success, but if it is not a life of virtue it may inspire little more than envy. On the other hand, a life which does not result in what the world considers "success," but a life lived virtuously, can inspire others in a multiplicity of ways.

Some look at us instrumentally, as a person who can be of use. Some look at us as an impediment, a person holding a position that they seek to have. Whenever our lives are successful in conventional ways we risk being used or envied. If our lives are also virtuous we have the opportunity to truly *inspire*. We have the opportunity to be seen as heroes and heroines. The same is true if our lives are less than successful but fully virtuous. As Aristotle said, you need not rule the earth and the sea to do good. Indeed, we often expect greater good from those who do not enjoy such conventional power or authority.

Mother Teresa had neither wealth nor titles. While we cannot all aspire to her level of accomplishment we can certainly emulate her virtues. The same is true for those around us. They may not be able to achieve what we have achieved; indeed, they may have achieved much more, in the conventional sense, but virtue can be appreciated, loved, and emulated wherever it is found. All of us can be heroes and heroines to others. In many cases we already are.

Appendix

HEROES/HEROINES SUGGESTED BY OUR STUDENTS (FROM FIGURE 3)

We invite readers to seek more detailed information on those individuals in the following list whose lives and experience interest them. They can then determine the degree to which these individuals had psychic wounds which they successfully transcended. Our students were not asked to suggest heroes/heroines with that framework in mind, but many will be found to fit that thematic template. Many of the details and some of the evaluative commentary here are drawn from *Wikipedia*, though facts have been checked with other sources.

Henry "Hank" Aaron (1934–) Baseball player. Nicknamed "Hammerin' Hank," his career spanned the years 1954–1976. Aaron grew up in poverty; his family could not afford standard baseball equipment and he practiced with bottle caps and sticks. His most notable achievement was setting the MLB record for most career home runs (755), a record which held for 33 years.

Dante Alighieri (1265–1321) Florentine poet. His *Commedia*, dubbed the *Divina Commedia* by Boccaccio, is considered the greatest work of Italian literature and a masterpiece of world literature. The poem describes Dante's journey through Hell, Purgatory and Paradise, guided initially by the poet Virgil and then by Dante's beloved Beatrice. Embroiled in the Guelph-Ghibelline conflict, Dante was exiled from Florence. A tomb was built for him in Florence in the basilica of Santa Croce, but his body remains in Ravenna.

Roald Engelbregt Gravning Amundsen (1872–1928) Norwegian explorer. Amundsen led the first Antarctic expedition (1910–1912) to reach the South Pole. The first person to reach both the North and South Poles, he is considered the first person to traverse the Northwest Passage. He disappeared in June 1928 while taking part in a rescue mission.

Marion Anderson (1897–1993) American contralto. The first African American to perform with the New York Metropolitan Opera, she debuted with the New York Philharmonic in 1925, singing at Carnegie Hall in 1928. Her European tours in the 1930's further established her reputation. In 1939 the Daughters of the American Revolution (DAR) refused permission for her to sing to an (integrated) audience in Constitution Hall. Thousands of DAR members, including First Lady Eleanor Roosevelt, resigned from the organization. She sang instead on the steps of the Lincoln Memorial before an audience of 75,000. She sang "My Country 'Tis of Thee."

Susan B. Anthony (1820–1906) Civil rights leader. Conquering her personal insecurities with regard to her looks and her public speaking abilities, she became a renowned leader in the struggle to secure women's suffrage in the United States.

Maria Corazón Sumulong Cojuangco-Aquino [Known as Corazón or Cory Aquino] (1933–2009) Political leader. After her husband, Sen. Benigno Aquino, Jr., a leading critic of the government of Ferdinand Marcos, was assassinated, Cory Aquino led the opposition against Marcos. With no prior political experience she unseated Marcos as President of the Philippines, serving from 1986–1992. Under her administration a new Constitution was enacted and several important legal reforms were adopted. She survived several attempted coups and dealt with major natural disasters besetting the Philippines.

Aristotle (384 B.C.–322 B.C.) Greek philosopher. A student of Plato and teacher of Alexander the Great, Aristotle wrote on many subjects, among them, physics, metaphysics, poetry, drama, music, rhetoric, logic, politics, ethics, biology and zoology. He made significant contributions to most of these subjects. His encyclopedic knowledge constitutes a significant portion of the foundation of western thought and his influence remains strong, several millennia after his death.

Neil Alden Armstrong (1930–) American astronaut. The first person to set foot on the Moon, Armstrong was a Navy pilot who saw extensive action in the Korean War. He uttered the famous words, "That's one small step for [a] man, one giant leap for mankind." The recipient of many awards and medals, there are more than a dozen United States elementary, middle and high schools named in his honor.

Arthur Robert Ashe, Jr. (1943–1993) Tennis player. Winner of three Grand Slam titles, he was the first African American ever selected to the United States Davis Cup Team. He served in the Army from 1966–68 and, in 1968, won both the United States Amateur Championship and U. S. Open. He remains the only African-American player to win the men's singles at Wimbledon, the U.S. Open, or Australian Open. Slowed by heart surgery in 1979, he retired in 1980. Contracting HIV through the blood transfusions received during one of his heart surgeries, he worked to call attention to the plight of AIDS sufferers worldwide. After his death his body lay in state at the Governor's Mansion in his home state of Virginia. The last individual to receive such an honor had been the Confederate general, Thomas 'Stonewall' Jackson.

Christiaan Neethling Barnard (1922–2001) Surgeon. A South African surgeon, Barnard was the first to perform a successful human-to-human heart transplant. He was also an outspoken opponent of South Africa's apartheid laws. After rheumatoid arthritis in his hands ended his surgical practice he established the Christiaan Barnard Foundation, which was dedicated to helping the world's underprivileged children.

David Robert Joseph Beckham, OBE (1975–) Footballer. Beckham plays midfield, is a member of the English national team and plays for the Los Angeles Galaxy. He has worked on behalf of UNICEF and serves as a spokesman for Malaria No More. In 2003 he was made an Officer in the Order of the British Empire by Queen Elizabeth II and in 2007 he was voted Britain's Greatest Ambassador.

Sir Timothy John Berners-Lee, OM, KBE, FRS, FREng, FRSA (1955–) Computer Scientist. He implemented the first successful communication between an HTTP client and server via the Internet and was ranked joint First in the *Daily Telegraph's* list of 100 greatest living geniuses. He is the director of the World Wide Web Consortium, which oversees the continued development of the WWW.

Margaret Bourke-White (1904–1971) Photographer and photojournalist. The first female war correspondent and the first woman to be allowed to work in combat zones during World War II, she was known to the *Life* magazine staff as 'Maggie the Indestructible'. Traveling through Germany with General Patton she photographed the appalling concentration camp, Buchenwald. She is quoted as saying, "Using a camera was almost a relief. It interposed a slight barrier between myself and the horror in front of me."

Rachel Louise Carson (1907–1964) Nature writer. A marine biologist, her writings are credited with advancing the global environmental movement. Her book *Silent Spring* (1962) called attention to the environmental results of synthetic pesticides and resulted in a nationwide ban on DDT and other chemicals. She was posthumously awarded the Presidential Medal of Freedom.

César Estrada Chávez (1927–1993) Labor leader and civil rights leader. An American farm worker of Mexican descent, he—with Dolores Huerta—co-founded the National Farm Workers Association, which eventually became the United Farm Workers. His birthday has become a holiday in eight U. S. states.

William Clark (1770–1838) Explorer, soldier and territorial governor. Clark led (with Meriwether Lewis) the Lewis and Clark Expedition of 1803–1805 across the Louisiana Purchase territory to the Pacific Ocean. Chiefly responsible for the drawing of maps, management of the expedition's supplies and hunting, he had previously served in the Army and later served as governor of the Missouri Territory. Multiple states have named counties in his honor and his name is now associated with various plants, fish and birds.

Nicolaus Copernicus (1473–1543) Scientist. The leading astronomer and one of the great polymaths of the renaissance, he was also a mathematician, physician, classicist, cleric, jurist, governor and diplomat. The first astronomer to formulate a scientifically-based heliocentric cosmology, displacing the Earth from the center of the universe, his *De revolutionibus orbium coelestium* is often regarded as both the starting point of modern astronomy and the beginning point for the Scientific Revolution.

Jacques Yves Cousteau (1910–1997) Explorer and filmmaker. The co-developer of the aqua-lung, he pioneered marine conservation. He produced more than 120 television documentaries, more than 50 books and an environmental protection foundation with 300,000 members. The recipient of numerous awards, including the Croix de guerre, he received the Presidential Medal of Freedom from President Reagan in 1985.

Crazy Horse [Lakota: Thašu ka Witko] (1842–1877) Warrior and leader. The war leader of the Oglala Lakota, he fought against the U.S. government in an effort to preserve the values and traditions of the Lakota people. He is being commemorated by the monumental sculpture of a mountain in the Black Hills of South Dakota, a short drive from Mt. Rushmore.

Marie Sklodowska-Curie (1867–1934) Scientist. A pioneer in the field of radioactivity, she was trained in both physics and chemistry and was the only person to have received Nobel Prizes in (these) two different sciences. Of Polish descent, she was also the first woman in France to complete a doctorate and the first female professor at the University of Paris. She intentionally refrained from patenting her radium-isolation process so that the scientific community could pursue this area of research without hindrance. During the First World War she pressed for use of mobile radiography units ("Little Curies") to treat wounded soldiers and, in her later years, headed a radioactivity laboratory and the Pasteur Institute at the University of Paris.

Charles Robert Darwin (1809–1882) Naturalist. An eminent collector and geologist, he proposed and provided scientific evidence that living species have evolved from common ancestors through the process he termed natural selection. With some modifications, his core discovery remains the foundation of biology. His book, *On the Origin of Species by Means of Natural Selection, or the Preservation of Favoured Races in the Struggle for Life* (1859) established evolution by common descent as the dominant scientific explanation of diversification in nature.

Michael Ellis DeBakey (1908–2008) Surgeon. A pioneer in the development of an artificial heart, DeBakey was the first surgeon to use an external heart pump successfully in a patient. This made open-heart surgery possible. In the course of his career he operated on more than 50,000 patients and practiced medicine until the day of his death at nearly 100 years of age. A renowned educator, innovator, and medical statesman, he received multiple awards including the Presidential Medal of Freedom and the National Medal of Science.

Bob Dylan [Robert Allen Zimmerman] (1941–) Musician. Singer, songwriter, author, poet and disk jockey, Dylan has been a dominant figure in popular music since the 1960's. Such songs as *Blowin' in the Wind* and *The Times They Are a-Changin'* helped define a generation. An inductee in the Rock and Roll Hall of Fame, his songs have received Grammy, Golden Globe and Academy Awards. *Rolling Stone* magazine ranked him as number two in its list of "Greatest Artists of All Time" (after the Beatles, above

Elvis Presley); he has received many other awards, including a Pulitzer Prize special citation.

Albert Einstein (1879–1955) Scientist. Winner of the Nobel Prize in physics, he is best known for his theory of relativity and his formula of mass-energy equivalence. Often considered the greatest scientist of the twentieth century, *Time* magazine named him the "Person of the Century" in 1999. His statement that "God does not play dice with the universe" has been frequently cited and analyzed.

Medgar Willy Evers (1925–1963) Civil rights activist. A soldier in the Second World War, Evers was instrumental in achieving the desegregation of the University of Mississippi and eventually studied law there, having previously been denied admission. Murdered by a member of the Ku Klux Klan, Byron De La Beckwith, he died in 1963. After two trials that failed to reach a verdict, Beckwith was brought to trial based on new evidence and was convicted in 1994, Medgar Evers' body having been exhumed for autopsy in the course of the trial. Beckwith died in prison in 2001.

Enrico Fermi (1901–1954) Scientist. An Italian physicist, Fermi is famous for his work on the development of the first nuclear reactor and for his contributions to other areas of physics, including quantum theory, nuclear and particle physics. He was awarded the Nobel Prize in physics in 1938 but died at the age of 53 of stomach cancer. Two of his graduate students who had worked with him on or near the nuclear pile also succumbed to death by cancer.

Henry Ford (1863–1947) Industrialist, philanthropist. Founder of the Ford Motor Company, he was the principal developer of the modern assembly line. His Model T automobile changed both American transportation and American industry. A prolific inventor, Ford was awarded 161 U.S. patents. He left the bulk of his wealth to the Ford Foundation, which reported assets of nearly $14 billion in 2007.

Annelies Marie "Anne" Frank (1929–1945) Diarist, holocaust victim. A German-born Jewish girl from Frankfurt, she is known for her diary, which chronicles her experiences hiding during the Nazi occupation of the Netherlands. She died of typhus in the Bergen-Belsen concentration camp, but her father survived and his efforts led to the publication of her diary in 1947. It was first published in English in 1952 as *The Diary of a Young Girl*.

Yuri Alexeyevich Gagarin (1934–1968) Cosmonaut. A member of the Soviet Air Force, Gagarin was the first human in space and the first to orbit the Earth. He received medals and plaudits from around the world for his pioneering work. He died during a routine training flight in 1968 near Kirzhach.

Galileo Galilei (1564–1642) Scientist. Born in the same year as William Shakespeare, Galileo was a Tuscan physicist, mathematician and astronomer. Sometimes termed the "father of modern science," his championing of Copernicanism brought him into conflict with the Catholic Church, which forced him to recant his heliocentrism and live out the last years of his life under house arrest. In 1992 Pope John Paul II expressed regret for the manner in which the Galileo affair was handled and officially acknowledged that the Earth was not stationary.

Mohandas Karamchand Gandhi (1869–1948) Political and spiritual leader. Known around the world as *Mahatma Gandhi* or 'great soul', he led the Indian independence movement and pioneered the resistance to tyranny through mass civil disobedience and non-violence. His birthday is commemorated in India as a national holiday.

William "Bill" Henry Gates, III (1955–) Industrialist, philanthropist. A self-professed college dropout and cofounder (with Paul Allen) of Microsoft, Gates is listed as the world's third richest person in 2008. Through the Bill and Melinda Gates Foundation, which was established in 2000, he has donated large amounts of money to multiple charitable organizations and scientific research programs, including a grant of $1 billion to the United Negro College Fund.

Henry Louis "Lou" Gehrig [b. Ludwig Heinrich Gehrig] (1903–1941) Baseball player. Known for his skill as a power hitter and for the longevity of his consecutive games played record (which stood for more than fifty years and earned him the nickname 'the iron horse'), Gehrig set other Major League records, including that for most career grand slam home runs (23) which still stands in 2008. He was voted the greatest first baseman of all time by the Baseball Writers' Association and chosen by fans in 1999 as the leading vote-getter on the Major League Baseball All-Century Team. The disease which claimed his life—amyotrophic lateral sclerosis—is now often termed 'Lou Gehrig's disease'.

Robert Frederick Xenon Geldof [Bob Geldof] (1951–) Musician and activist. Irish singer, songwriter and actor, Geldof achieved fame as a member of the rock band, The Boomtown Rats. He has since become known for his philanthropic work, which includes the organizing of the Live Aid and Live 8 concerts.

John Herschel Glenn, Jr. (1921–) Astronaut and politician. A Marine Corps fighter pilot, Glenn joined the *Mercury Seven*, NASA's original astronaut group. The first American to orbit the Earth, he was also the oldest person to fly in space, joining the crew of the Space Shuttle *Discovery* in 1998 at the age of 77. He served as a senator from Ohio from 1974–1999. In 1990 he was inducted into the Astronauts Hall of Fame.

William Franklin Graham, Jr. [Billy Graham] (1918–) Evangelist. Spiritual adviser to multiple U.S. presidents, he has preached in person to more people around the world than anyone who has ever lived. He was ranked seventh on Gallup's list of admired people for the twentieth century. A vocal opponent of segregation in the 1960's he was also a staunch critic of apartheid in South Africa. During the Cold War he was the first evangelist of note to speak behind the Iron Curtain, calling for peace.

Dag Hjalmar Agne Carl Hammarskjöld (1905–1961) Diplomat. The second Secretary-General of the United Nations, he sought to address problems and settle crises on multiple fronts, serving from 1953 until his death in a plane crash in 1961. He is the only individual to have been awarded the Nobel Peace Prize posthumously.

Stephen William Hawking (1942–) Scientist. A theoretical physicist, Hawking held the Lucasian Professorship of Mathematics at Cambridge for thirty years until his retirement in 2009, a position once held by Isaac Newton. He is known for his studies of black holes, his work in cosmology and quantum gravity and the popular works he has written on these subjects. He suffers from amyotrophic lateral sclerosis, but continues unabated in his scientific research.

Sir Edmund Hillary (1919–2008) Explorer. A New Zealander, Hillary, with Sherpa mountaineer Tenzing Norgay, was the first to reach the summit of Mount Everest (on May 29, 1953 at the age of 33). Following this accomplishment he devoted part of the remainder of his life to helping the Sherpa people of Nepal. Through his efforts many schools and hospitals have been built in this part of the Himalayas.

Íñigo Oñaz López de Loyola [Saint Ignatius of Loyola] (1491–1556) Religious leader. The principal founder and first Superior General of the Society of Jesus, he was particularly instrumental in the founding of schools, colleges and seminaries. Jesuit institutions now exist throughout the world. He is also known for his *Spiritual Exercises*, which form the basis for Jesuit religious retreats.

Thomas Jefferson (1743–1826) President. The third President of the United States and principal author of the Declaration of Independence, Jefferson was one of the most influential Founding Fathers. An architect, inventor, horticulturist, and author, he founded the University of Virginia. As President he acquired the Louisiana Purchase territory and sponsored the Lewis and Clark Expedition. He is honored by a substantial monument in Washington and his face is featured on Mt. Rushmore.

Joan of Arc (1412–1431) Saint and soldier. Also known as the "Maid of Orleans," she led the French army to a number of significant victories in the Hundred Years' War. Claiming divine guidance, she was eventually captured by the English and burned at the stake at the age of 19. She was beatified in 1909 and canonized in 1920.

Pope John XXIII [Angelo Giuseppe Roncalli] (1881–1963) Roman Pontiff. Praised for his ecumenism and personal spirit by members of many churches, Pope John XXIII was beatified in 2000. The 261st Pope of the Roman Catholic Church, he is known for his outgoing, personal warmth, his travels beyond the confines of the Vatican and his calling for the Second Vatican Council, which reshaped Catholicism.

Pope John Paul II [Karol Józef Wojtyla] (1920–2005) Roman Pontiff. The 263rd Pope of the Roman Catholic Church, he traveled far more extensively than any of his predecessors, visiting over 100 countries and utilizing the multiple languages in which he was fluent. In 1981 he was critically wounded by a Turkish gunman, lost nearly three-quarters of his blood, and underwent five hours of surgery. He later visited his would-be assassin in prison and told him that he had forgiven him. He attributed his survival to the help of Our Lady of Fatima. Considered too conservative by some, he markedly extended the influence of the Church in the Third World.

Samuel Johnson (1709–1784) English writer and lexicographer. The preeminent literary figure of the English eighteenth century, he excelled in multiple genres and accomplished a series of major editorial projects, including his edition of the works of Shakespeare and his compilation of the first, great English dictionary. Born in modest circumstances in Staffordshire, he attended Oxford for a year (leaving because of a lack of funds) and eventually came to London with an unfinished play and the hope to pursue writing for hire until he could become established. Eventually he gave his name to an age. Devoutly religious and politically conservative, he was a great advocate of the new science and a fervent opponent of slavery and colonialism, a supporter of the poor (to whom he gave approximately two-thirds of his pension) and of his poor friends, to whom he offered shelter within his home. His black manservant, Francis Barber, was the residuary legatee of his estate. He is honored by his birthplace museum in Lichfield, the museum at his Gough Square Residence in London (where he compiled his dictionary) and by a memorial in Westminster Abbey, where he is buried.

Chief Joseph [b. Hinmuuttu-yalatlat (alternatively Hinmaton-Yalaktit or Hin-mah-too-yah-lat-kekt, "Thunder Rolling Down the Mountain")] (1840–1904) Chief and moral leader. Chief of the Wal-lam-wat-kain (Wallowa) band of Nez Perce Indians at the time of General Oliver O. Howard's attempt to forcibly remove his people and other "non-treaty" Indians to a reservation in Idaho. He is renowned as a humanitarian and peacemaker for his resistance to the removal. He spoke eloquently against the injustice of U.S. policy toward his people and hoped for a time in which America's promise of freedom and equality would be extended to Native Americans as well. He died in exile from his homeland, according to his doctor "of a broken heart."

Jan Karski [Jan Kozielewski] (1914–2000) Patriot, soldier and political scientist. A Polish resistance fighter during the Second World War, he was captured by both the Red Army and the Gestapo. Brutally tortured, his entire face needed to be rebuilt. He repeatedly crossed enemy lines to act as a courier between his occupied country and the West. He observed the conditions in the Warsaw Ghetto and saw the mass murders in a Nazi concentration camp in Eastern Poland. In a 1943 meeting with President Roosevelt, he personally apprised him of the nature and dimensions of the holocaust. He later served as a Professor of Government at Georgetown University. A Roman Catholic, he was made an honorary citizen of Israel in 1994 and received the highest decorations of the Polish government. In Jerusalem a tree was planted in his name in the Alley of the Righteous Among the Nations. The 'Jan Karski Corner' was established in New York City at the corner of 37th Street and Madison Avenue, where a statue commemorates his courage, integrity and great personal dignity.

Helen Adams Keller (1880–1968) Author, activist and lecturer. The first deaf/blind person to graduate from college, she became a prolific author, traveler, and spokesperson for progressive causes. She was awarded the Presidential Medal of Freedom by President Johnson and elected to the Women's Hall of Fame at the New York World's Fair in 1965.

Martin Luther King, Jr. (1929–1968) Civil rights leader. A Baptist minister and the first president of the Southern Christian Leadership Conference, his efforts led to the 1963 March on Washington, where he delivered his celebrated "I Have a Dream" speech. A great orator, he was the youngest person to receive the Nobel Peace Prize. He was assassinated on April 4, 1968 in Memphis and was posthumously awarded the Presidential Medal of Freedom and the Congressional Gold Medal. President Reagan signed a bill in 1983 creating a federal holiday to honor him. Martin Luther King, Jr. Day was established as a national holiday in 1986. According to a Gallup poll, Dr. King is the second most admired person in the 20th century; the first is Mother Teresa.

General Robert E. Lee (1807–1870) Military leader. A career military officer, Lee was among the most celebrated generals in American history. He declined the command of the Union Army because of his loyalty to his home state of Virginia and eventually served as a senior military adviser to President Jefferson Davis and commanded the Army of Northern Virginia. A superb tactician, he continued to be honored after the defeat of the Confederacy. After the war he served as president of Washington College, whose name was changed to Washington and Lee University after his death.

Meriwether Lewis (1774–1809) Explorer, soldier and public administrator. With William Clark he led the Corps of Discovery, the Lewis and Clark Expedition. President Jefferson spoke of his "undaunted courage" and his "firmness and perseverance of purpose." Various flora and fauna are named for him and he is recognized for contributions to science that have been of immense (some say incalculable) importance. Ft. Lewis, Washington is named in his honor.

Abraham Lincoln (1809–1865) President. The 16th President of the United States, he led the country through its gravest internal crisis, the Civil War. Assassinated less than a week after the war's end, he was seen by many as a martyr. He helped preserve the Union and introduced measures which resulted in the abolition of slavery. Though often vilified in his time he is now considered by many to be the greatest of all American presidents. His face is carved on Mt. Rushmore and his monument in Washington is one of the most frequently-visited and most admired.

Martin Luther (1483–1546) Reformer and spiritual leader. A monk, theologian, and university professor, he is considered the father of Protestantism. His role in and influence on the Protestant Reformation changed Western history. His theology challenged the authority of Rome and has been immensely influential. His translation of the Bible into the vernacular made the scriptures more accessible to lay readers and influenced both the church and German culture. His hymns continue to be sung today and Lutheran Churches draw a congregation of over sixty million.

Sharon Christa Corrigan McAuliffe (1948–1986) Teacher and astronaut. A teacher from New Hampshire, she was selected from a list of more than 11,000 applicants to participate in the NASA Teacher in Space Project. With six other crew members she was killed in the Space Shuttle Challenger disaster. She was posthumously awarded the Congressional Space Medal of Honor.

Máiread Corrigan Maguire (1944–) Peace activist. She is the co-founder, with Betty Williams, of the Community of Peace People, an organization dedicated to peacefully resolving the Troubles in Northern Ireland. She became active in the peace movement when three children of her sister, Anne Maguire, were killed by a car driven by a Provisional Irish Republican Army man who was fatally shot by British troops while attempting to escape. Her sister later committed suicide. She and Betty Williams were co-recipients of the Nobel Peace Prize in 1976.

Nelson Rolihlahia Mandela (1918–) Political and moral leader. A former President of South Africa, the first to be elected in a fully representative democratic election, he was previously an anti-apartheid activist and leader of the African National Congress. Imprisoned for 27 years, he credited Mahatma Gandhi for inspiring in him a philosophy of non-violence and a determination to face adversity with dignity. He received the Nobel Peace Prize in 1993.

Beryl Clutterbuck Markham (1902–1986) Pilot, author and adventurer. A British-born Kenyan author, she is principally remembered for her east-to-west non-stop Atlantic crossing. When her fuel tank vents iced she was forced to crash-land near Cape Breton, Nova Scotia, but she was still the first woman to accomplish this crossing. Her adventures are recounted in her memoir, *West With The Night.*

Michelangelo [Michelangelo di Lodovico Buonarroti Simoni] (1475–1564) Artist and quintessential Renaissance man. Painter, sculptor, architect, poet and engineer, he made significant contributions in each of these fields. Among his many important works are the *Pietà*, the *David* and the paintings on the ceiling and altar wall of the Sistine Chapel in Rome. He also designed the dome of St. Peter's Basilica there. He was the first Western artist whose biography was published during his own lifetime.

Mother Teresa [Nënë Tereza, Agnes Gonxhe Bojaxhiu] (1910–1997) Nun and minister to the poor and sick. An Albanian-born Roman Catholic nun and Indian citizen, she founded the Missionaries of Charity in Calcutta in 1950. She is internationally known as a humanitarian and minister to the poor, the helpless and the infirm. Her Missionaries of Charity now include over 4,000 sisters and 100,000 lay volunteers operating 610 missions in 123 countries. She once said that what all of those to whom she ministered needed was, simply, love. She received the Nobel Peace Prize in 1979 and was beatified by Pope John Paul II.

First Lieutenant Audie Leon Murphy (1924–1971) Soldier and actor. The most decorated United States combat soldier in United States military history, he received the Medal of Honor and 32 additional U.S. medals, five from France and one from Belgium. Forced to place his three younger siblings in an orphanage during his absence, he reclaimed them after the war. He later became an actor, appearing in 44 films. He died in a plane crash in 1971 and was interred, with full military honors, in Arlington National Cemetery. His gravesite is the second most-visited, after that of President John F. Kennedy.

Vice-Admiral Horatio Nelson, 1st Viscount Nelson (1758–1805) Naval commander. He is particularly famous for his efforts in the Napoleonic Wars, most notably in the Battle of Trafalgar, a decisive British victory in which he lost his life. (His ship, the *Victory*, was towed to Gibraltar, his body being preserved in a cask of brandy. The bullet that killed him is now on public display at Windsor Castle.) He was an exceptional commander, perhaps the greatest naval commander in human history. Historians particularly cite his ability to inspire officers of the highest rank and able-bodied seamen of the lowest. Among his many memorials, the most notable is Nelson's Column, which stands at the center of Trafalgar Square.

Sir Isaac Newton (1643–1727) Scientist. Creator (with Leibniz) of the calculus, he was a physicist and astronomer who did significant work in optics. He is known especially for his *Philosophiae Naturalis Principia Mathematica* (1687) in which he describes both universal gravitation and his three laws of motion, the groundwork for classical mechanics, which dominated the scientific view of the physical universe for three centuries.

Florence Nightingale (1820–1910) Nurse, writer and statistician. Born into a rich British family, living in a Florentine villa, she was named for the city of her birth. She alienated her family, particularly her mother, by electing to become a nurse, at that time a career pursued by poorer women. Best known for her work during the Crimean War, where she realized the drastic effects on health of poor living conditions, she became a great advocate for the amelioration of those conditions, both in the military and in civilian hospitals. According to the BBC she became the most famous woman in Britain after Queen Victoria. She made extensive use of statistics and graphics in her analyses of health conditions and studies of epidemiology and made special contributions to the improvement of medical care in India. She was elected the first female member of the Royal Statistical Society and later became an honorary member of the American Statistical Association. Her most important contribution was her role in founding the modern profession of nursing. A statue of her stands in Waterloo Place, London.

Tenzing Norgay (1914–1986) Mountaineer. A Nepalese/Tibetan mountaineer, he is often referred to as Sherpa Tenzing. On May 29, 1953, he and Sir Edmund Hillary were the first to reach the summit of Mount Everest. They remained friends throughout their lives. Though Tenzing never learned to read or write, he spoke several languages. He died of a cerebral hemorrhage in India in 1986.

Sandra Day O'Connor (1930–) Jurist. The first woman to serve on the Supreme Court, she did so from 1981 until her retirement from the bench in 2005. In spite of her many accomplishments as a law student, no California firm would hire her, though one offered her a position as a legal secretary. In her first year on the Supreme Court, to which she was appointed by President Reagan, she received over 60,000 letters from the public, more than any other justice in previous history. She stepped down from the court in order to spend more time with her ill husband. The federal courthouse in Phoenix is named in her honor.

James Cleveland "Jesse" Owens (1913–1980) Athlete. He participated in the 1936 Summer Olympics in Berlin, where he achieved international fame by winning four gold medals, undercutting Hitler's pretensions with regard to Aryan racial superiority. He was the grandson of a slave and the son of a sharecropper. Known as the "Buckeye Bullet," he won eight individual

NCAA championships as a student at Ohio State University. He was honored by Presidents Eisenhower, Ford and George H. W. Bush. A street in Berlin was named for him and the Ohio State University track and field facility is named the Jesse Owens Memorial Stadium.

Rosa Louise McCauley Parks (1913–2005) Civil rights activist. Termed the "Mother of the Modern-Day Civil Rights Movement," she refused to obey a bus driver's order that she give up her seat to make room for a white passenger. Her action (on December 1, 1955) sparked the Montgomery Bus Boycott and turned her into an international symbol of resistance to racial segregation. President Clinton awarded her the Presidential Medal of Freedom. *Time* magazine named her one of the 20 most influential figures of the twentieth century. President George W. Bush issued a proclamation ordering that all flags on U.S. public areas be flown at half-staff on the day of her funeral. The bus on which she was riding is now on exhibit at the Henry Ford Museum in Dearborn, Michigan.

Linus Carl Pauling (1901–1994) Scientist and peace activist. One of only two people to receive Nobel prizes in two fields (the other is Madame Curie), he received the prize for Peace and for Chemistry. Sometimes termed the "father of molecular biology" he made pivotal discoveries with regard to sickle cell anemia. He was also a vocal advocate of the importance of vitamins and minerals in the prevention of disease. In 2008 the United States Postal Service released a 41 cent stamp honoring him and he was inducted into the California Hall of Fame in December, 2008.

Pelé [Edson Arantes do Nascimento] (1940–) Footballer. The top scorer in the history of the Brazil national team and the only footballer to be a part of three World Cup-winning teams. He was proclaimed the Athlete of the Century by the International Olympic Committee. He is a member of the American National Soccer Hall of Fame, has received an honorary British knighthood, and, in 2005, won the BBC Sports Personality of the Year Lifetime Achievement Award. A UNESCO Goodwill Ambassador, he is generally considered the greatest footballer of all time.

Plato (429/423 B.C.–348/347 B.C.) Greek philosopher. Founder of the Academy in Athens—the first institution of higher learning in the Western world—he was a student of Socrates and the teacher of Aristotle. His work represents a pivotal portion of the foundation of Western philosophy. Alfred North Whitehead once said that, "The safest general characterization of the European philosophical tradition is that it consists of a series of footnotes to Plato."

Ronald Wilson Reagan (1911–2004) President and former actor. The 40th President of the United States, he previously served as Governor of California after a career as an actor and spokesperson for General Electric. Born in Illinois he moved to California in the 1930's and served as president of the Screen Actors Guild. As President he reduced federal income tax rates and played a role (some argue a significant one) in ending the Cold War. Surviving an assassination attempt in 1981, he was noted for his sense of humor and skills in communication. He was sometimes termed "the great communicator." A Gallup poll in 2001 ranked him as the greatest President in U.S. history, while one in 2007 ranked him second after Abraham Lincoln. He died in 2004 after an extended bout with Alzheimer's Disease. He was the 10th President to lie in state after his death. During the 34 hours in which his body lay there, nearly 105,000 people filed past the coffin.

Jackie Robinson (1919–1972) Athlete. The first African-American major league baseball player of the modern era, he played for the Brooklyn Dodgers, beginning in 1947, when he won the first Rookie of the Year award. This ended approximately eighty years of segregation in baseball. While sometimes being subjected to fan hostility and insults, he played with great dignity and with uncommon skill. Branch Rickey, the owner of the Dodgers, has received praise for his courage in signing him. He was inducted into the Baseball Hall of Fame in his first year of eligibility, 1962. A political activist after his retirement from baseball, he was posthumously awarded the Presidential Medal of Freedom.

Franklin Delano Roosevelt (1882–1945) President. The 32nd President of the United States he was often referred to as FDR. Elected to four terms, serving from 1933–1945, he is the only President to serve more than two terms. Leading the country through both the Great Depression of the 1930's and the Second World War, he is often considered one of the country's greatest presidents. He labored with the effects of polio throughout his life and he created the New Deal to address the plight of the unemployed, support the recovery of the economy and reform the economic and banking systems. One of his most frequently cited legacies is the U. S. Social Security system. According to a Gallup poll of U.S. citizens, he was the sixth most admired person of the 20th century.

Theodore Roosevelt (1858–1919) President. The 26th President of the United States he was often referred to as T.R. A Governor of New York, a soldier, historian and naturalist, he is renowned for his energy, his larger-than-life personality and broad range of interests. He is the only individual to hold both the Nobel Prize for Peace and the Medal of Honor (awarded posthumously for his actions in the Spanish-American War). Famous for his trust-busting and his promotion of the conservation movement, he considered his principal international achievement to be the construction of the Panama Canal. Consistently ranked among our greatest presidents, his image stands on Mt. Rushmore.

Wilma Glodean Rudolph (1940–1994) Athlete. She was the first American woman to win three gold medals in track and field during a single Olympic Games, though she was running on a sprained ankle. An African American and the twentieth in a family of twenty-two children, she was diagnosed with polio at the age of 4 and her mother Blanche was required to drive her from their home in Clarksville, Tennessee to Fisk University's Meharry Hospital in Nashville, a hospital for African Americans, fifty miles from their home, twice each week. With the help of her parents and siblings she was able to overcome her disease and become one of the greatest athletes in U.S. history, a member of the U.S. Olympic Hall of Fame. Her homecoming parade and the banquet that followed that evening, were the first such events in Clarksville to be integrated. Later in her life she worked as a teacher and coach, succumbing to brain cancer at the age of 54. She is also remembered as one of the celebrated "Tigerbells," Coach Ed Temple's women's track team at Tennessee State University.

Andrei Dmitrievich Sakharov (1921–1989) Scientist and human rights activist. An eminent nuclear physicist, he was a vocal advocate for civil liberties and reforms in the Soviet Union. Awarded the Nobel Peace Prize in 1975, he was not permitted to leave the Soviet Union to collect it. After years of surveillance and harassment he was finally permitted to return to Moscow when Mikhail Gorbachev initiated the policies of *perestroika* and *glasnost.* In 1989 he was elected to the new parliament, the All-Union Congress of People's Deputies and co-led the democratic opposition. He died of a heart attack in 1989 at the age of 68.

Jonas Edward Salk (1914–1995) Scientist and physician. The vaccine which he developed was instrumental in the eradication of polio. Once the 'Salk vaccine' was available, the number of cases in the U.S. dropped by 85–90% in two years. In 1965 he established the Salk Institute for Biological Studies in La Jolla, California, where the major research emphases were molecular biology and genetics. He directed the institute until his retirement in 1985. He was awarded the Presidential Medal of Freedom by President Carter in 1977. Once when asked by Edward R. Murrow concerning the ownership of the vaccine patent, he responded, “There is no patent. Could you patent the sun?”

Albert Schweitzer (1875–1965) Physician. A distinguished theologian, musician and philosopher as well as a physician, he received the 1952 Nobel Peace Prize for his philosophy of “reverence for life”, expressed in many ways, but most famously in founding and sustaining the Albert Schweitzer Hospital in Lambaréné, now in Gabon, west central Africa (then French Equatorial Africa). After briefly utilizing a shed that had been a chicken house, he and his wife built their first hospital of corrugated iron in 1913. After the war he returned there and constructed new hospital buildings. From the receipt of the Nobel Peace Prize in 1952 until his death he worked against nuclear testing and nuclear weaponry with Albert Einstein and Bertrand Russell. He died in 1965 at his hospital in Gabon. His grave, on the banks of the local river, is marked by a cross which he made himself.

Sir Ernest Henry Shackleton (1874–1922) Explorer. He participated in four Antarctic expeditions, three of which he led. He is best known for leading the unsuccessful Imperial Trans-Antarctic Expedition, sometimes known as the “Endurance Expedition” (1914–1916). When his ship was trapped in the ice and destroyed, he led his men to refuge on Elephant Island. He then sailed in an open boat with five other men to South Georgia Island, a distance of 800 miles. The seas were high and the lead-gray skies permitted very few navigational sightings. Arriving at the Island, they were forced to cross a mountainous region in order to reach the local whaling station. He was eventually able to rescue all of his men left on Elephant Island. He is buried near the former whaling station at Grytviken on South Georgia. A statue of Shackleton stands outside the headquarters of the Royal Geographical Society in London.

Socrates (469 B.C.-399 B.C.) Greek philosopher. A foundational figure in Western philosophy, his thought survives through the dialogues of his student, Plato. He is principally known for his dialectical method of inquiry, now often termed the “Socratic Method.” He was found guilty of corrupting the minds of the young in Athens and he was sentenced to death by drinking a mixture that contained poison hemlock. His death is described in Plato’s *Phaedo*.

Vice-Admiral James Stockdale (1923–2005) Naval hero. He is one of the most highly decorated Naval officers in American history. The highest ranking officer held as a prisoner during the Vietnam war, he was determined to frustrate enemy attempts to use him for propaganda purposes and injured himself so that he could not be filmed or photographed. He was later unable to return to active flying status because of the results of torture and other maltreatment. He served as President of the Naval War College and President of the Citadel. He eventually became a fellow of the Hoover Institution at Stanford University. He was awarded 26 personal combat decorations, including four Silver Stars and the Medal of Honor. He ran for the vice presidency on Ross Perot's independent ticket in 1992 and died in 2005. In 2006 the Navy announced that the USS Stockdale, a guided missile destroyer, would be named for him. It was christened in 2008. The United States Naval Academy now commemorates him with the Vice Admiral James B. Stockdale Center for Ethical Leadership.

Captain Chesley "Sully" Sullenberger (1951–). Airline transport pilot, accident investigator and safety expert. On January 15, 2009 he was piloting an Airbus A320 (US Airways Flight 1549/United Airlines Flight 1919) from LaGuardia Airport to Charlotte/Douglas International Airport when the plane was struck by a large flock of birds which disabled both of its engines. Realizing that a conventional landing was then impossible, he successfully ditched in the Hudson River. All crew members and all passengers (a total of 155 individuals) survived. Both the US Senate and House of Representatives passed resolutions honoring him, his co-pilot Jeff Skiles, the crew, passengers and first responders to the emergency landing. He has recounted the events of that day in his book *Highest Duty: My Search for What Really Matters* (2009). (A contemporary example of heroism, added by the authors.)

Corrie ten Boom (1892–1983) Holocaust survivor. A Dutch Christian, she helped many Jews escape from the Nazis during the Second World War. The Germans arrested her entire family with the help of a Dutch informant and several did not survive. She was released on Christmas Day in 1944 and, after the war, returned to the Netherlands to set up rehabilitation centers. A prolific author, her most famous book is *The Hiding Place* (1971), which was made into a film. In 1967 she was honored as one of the Righteous Among the Nations by the State of Israel. She was knighted by the Queen of the Netherlands in recognition of her work during the war and a museum in the Dutch city of Haarlem is dedicated to her and her family.

Margaret Hilda Thatcher, Baroness Thatcher (1925–) British Prime Minister. Leader of the Conservative Party from 1975–1990 and Prime Minister from 1979–1990, she is the first and only woman to date to hold either position. Her opposition to the Soviet Union and strong rhetoric earned her the nickname the "Iron Lady." She read Chemistry at Oxford and served there as the President of the Oxford University Conservative Association. As Prime

Minister she was closely allied with President Reagan and persuaded him of Mikhail Gorbachev's sincere desire to reform the Soviet Union. She narrowly escaped the 1984 Brighton hotel bombing by the IRA during the Conservative Party Conference and insisted that the conference open promptly the next day in defiance of the bombers, an act that was widely approved across the political spectrum. In 2007 she became the first Prime Minister to be honored with a statue in the House of Commons while still alive. The bronze statue stands opposite that of her political hero, Winston Churchill.

Jacobus Franciscus "Jim" Thorpe (1888?-1953) Athlete. An athlete of incomparable versatility, he won Olympic gold medals in both the pentathlon and decathlon, played football in college and professionally and also played professional baseball and basketball. He was later stripped of his Olympic medals when it was discovered that he was paid for playing minor league baseball, thus violating the amateur status rules. It was believed that this action may have resulted from ethnic prejudice, since he was of mixed Native American and white ancestry. He was named the greatest athlete of the first half of the twentieth century by the Associated Press and ranked third on their list of athletes of the century in 1999. When his sports career ended he lived in poverty and failing health. His medals were restored in 1983, thirty years after his death.

Desmond Mpilo Tutu (1931–) Cleric and activist. The first black South African Anglican Archbishop of Cape Town and primate of the Church of the Province of Southern Africa, he rose to worldwide fame during the 1980's as a vocal opponent of apartheid. Termed "South Africa's moral conscience," he campaigns to fight AIDS, poverty and racism. He received the Nobel Peace Prize in 1984, the Albert Schweitzer Prize for Humanitarianism and the Gandhi Peace Prize in 2007.

Lech Wałęsa (1943–) Trade unionist and political leader. Co-founder of *Solidarność*, the Soviet bloc's first independent trade union, he lost his job in the Lenin Shipyard of Gdańsk in 1976. In 1981 he was arrested and sent to southeastern Poland near the Soviet border for 11 months. In 1983 he was awarded the Nobel Peace Prize and in 1990 he won the presidential election, serving as President of Poland from 1990 to 1995. Recipient of 32 honorary degrees from American and European universities, he was named "Man of the Year" by *The Financial Times* (1980), *The Observer* (1980) and by *Time* magazine (1981).

Raoul Gustav Wallenberg (1912–1947?) Humanitarian. A Swedish national, he worked to save the lives of Hungarian Jews from being sent to concentration camps by issuing them protective passports from the Swedish embassy. He has been credited with saving 15,000 lives. In 1945 he was arrested on the direct order of Soviet Deputy Commissar for Defense Nikolai Bulganin, though the ultimate order may well have come from Stalin. Responding to diplomatic pressure, the Soviets announced that he had died of a heart attack in 1947 in Lubyanka prison in Moscow. After a study of classified archival documents, the deputy head of the KGB intelligence service later concluded that Wallenberg had been shot. The State of Israel has honored him as one of the Righteous Among the Nations and President Reagan signed the legislation in 1981, making him an Honorary Citizen of the United States, the second person to receive this honor, the first being Winston Churchill, who, unlike Wallenberg, had one U.S. parent. A monument commemorating him stands opposite the headquarters of the United Nations in New York.

George Washington (1732–1799) President. Sometimes termed the country's "father," he was the commander of the Continental Army and the first President of the United States. In his funeral oration, Henry "Light Horse Harry" Lee (Robert E. Lee's father) said that Washington was "first in war, first in peace, and first in the hearts of his countrymen." His picture appears on the $1 bill; his image is carved on Mt. Rushmore and he is generally considered to be one of the greatest of the American presidents. His life is the subject of legend and lore; his home in Mount Vernon is visited by approximately one million people each year and the 555+ foot obelisk erected in his memory is one of the most prominent elements of the Washington skyline.

Eliezer "Elie" Wiesel (1928–) Writer and holocaust survivor. Awarded the Nobel Peace Prize in 1986, he has been an advocate for a multiplicity of causes and has spoken out forcibly and eloquently against violence, repression and racism. Most recently he has voiced support for intervention in Darfur. The author of over forty books, the best known is his memoir, *Night*, which describes his experiences during the holocaust.

Theodore Samuel "Ted" Williams (1918–2002) Baseball player and Marine pilot. One of the greatest hitters (if not the greatest) in baseball history, he played 19 seasons for the Boston Red Sox, his career being interrupted on two occasions for military service in the Second World War and in Korea. He had a career batting average of .344 and his .551 on base percentage record stood for 61 years. He was inducted into the Baseball Hall of Fame in 1966. In his induction speech he included a statement calling for the recognition of the great players from the Negro Leagues. On his last at bat on September 28, 1960 he hit a home run.

Thomas Woodrow Wilson (1856–1924) President. The twenty-eighth President of the United States, he also served as Governor of New Jersey and President of Princeton University. He worked with a Democratic Congress to create the Federal Trade Commission, pass the Clayton Antitrust Act and create the Federal Reserve System. He did not learn to read until he was 12 years old but succeeded academically through determination and great self-discipline. Largely for his efforts to create the League of Nations, he was awarded the Nobel Peace Prize in 1919.

Eldrick Tont "Tiger" Woods (1975–) Golfer. He has more career major wins and career PGA Tour wins than any other active golfer, is the youngest player to achieve the career Grand Slam, and the youngest and fastest to win 50 tournaments on tour. Named AP Male Athlete of the Year a record-tying four times and the only person to be named *Sports Illustrated's* Sportsman of the Year more than once, he is on track to be the first athlete to pass one billion dollars in earnings. Multiracial, he has literally helped to change the face of golf.

Frank Lloyd Wright (1867–1959) Architect. Both an architect and noted interior designer, writer and educator, he designed more than 1,000 projects, of which more than 500 were actually completed. His residential home in southwestern Pennsylvania, *Fallingwater,* is sometimes termed the most important architectural work of the twentieth century. The leader of the Prairie School architectural movement, his *oeuvre* includes a wide array of building types. In 1991 the American Institute of Architects recognized him as "the greatest American architect of all time."

Wilbur (1867–1912) and Orville (1871–1948) Wright Aircraft engineers. They are generally credited with inventing and building the first successful airplane and making the first controlled, powered and sustained heavier-than-air human flight on December 17, 1903. In the two years following they developed their flying machine into the first practical fixed-wing aircraft. They are recognized in a special exhibit at the National Air and Space Museum.

Alvin Cullum "Sergeant" York (1887–1964) Soldier. A soldier in the First World War, he led an attack on a German machine gun nest, taking 32 machine guns, killing 28 German soldiers and capturing 132 others in the Meuse-Argonne Offensive. The Distinguished Service Cross which he received was later upgraded to the Medal of Honor and was presented to him by

General John J. Pershing, the commanding general of the American Expeditionary Force. He received other awards from France, Italy and Montenegro. His story was told in the 1941 film *Sergeant York*. He refused to authorize a film version of his life unless he received a contractual guarantee that he would be portrayed by Gary Cooper. For his performance, Cooper received the Academy Award for Best Actor.

Bibliography

Aldington, Richard. (1955). *Lawrence of Arabia: A Biographical Enquiry*. London: Collins.

Alexander, Caroline. (1998). *The Endurance: Shackleton's Legendary Antarctic Expedition*. New York: Knopf.

Ambrose, Stephen E. (1975). *Crazy Horse and Custer: The Parallel Lives of Two American Warriors*. New York: Doubleday.

———. (1998). *The Victors: Eisenhower and His Boys, The Men of World War II*. New York: Simon and Schuster.

Aristotle. (1941). *The Basic Works*. Ed. Richard McKeon. New York: Random House.

———. (1948). *On the Art of Poetry*. trans. S. H. Butcher. Indianapolis: Bobbs-Merrill.

Auerbach, Erich. (1953). *Mimesis: The Representation of Reality in Western Literature*. Princeton, NJ: Princeton University Press.

Austen, Jane. (1988). *The Oxford Illustrated Jane Austen*. Ed. R. W. Chapman, 3rd ed., 6 vols. Oxford: Oxford University Press.

Ayer, A. J. (1985). *Wittgenstein*. London: Weidenfeld and Nicolson.

———. (1990). *The Meaning of Life*. New York: Scribner's.

Baldwin, A. L. (1967). *Theories of Child Development*. 2nd. ed. New York: Wiley.

Bandura, A. (1977). *Social Learning Theory*. Englewood Cliffs, NJ: Prentice-Hall.

———. (1986). *Social Foundations of Thought and Action*. Englewood Cliffs, NJ: Prentice-Hall.

Bandura, A. and R. H. Walters. (1963). *Social Learning and Personality Development*. New York: Holt.

Barbash, Tom. (2003). *On Top of the World: Cantor Fitzgerald, Howard Lutnick, and 9/11: A Story of Loss and Renewal*. New York: HarperCollins.

Barczewski, Stephanie. (2007). *Antarctic Destinies: Scott, Shackleton and the Changing Face of Heroism*. London: Hambledon Continuum.

Barnard, Ellsworth, ed. (1944). *Shelley: Selected Poems, Essays, and Letters.* New York: Odyssey.

Bartlett, Robert A. (2001). *The Karluk's Last Voyage.* New York: Cooper Square Press.

Barzun, Jacques. (1989). *The Culture We Deserve.* Arthur Krystal, Ed. Middletown, CT: Wesleyan University Press.

Bate, Walter Jackson. (1955). *The Achievement of Samuel Johnson.* New York: Oxford University Press.

———. (1975). *Samuel Johnson.* New York: Harcourt, Brace, Jovanovich.

Bateson, Mary Catherine. (1989). *Composing A Life.* New York: Plume.

Beccaria, Cesare. (1963). *On Crimes and Punishments* (1764). Trans. Henry Paolucci. Indianapolis: Bobbs-Merrill.

Benjamin, Walter. (1969). "The Work of Art in the Age of Mechanical Reproduction," (1936), in *Illuminations*, trans. Harry Zohn. New York: Schocken.

Bennett, William J., ed. (1993). *The Book of Virtues: A Treasury of Great Moral Stories.* New York: Simon and Schuster.

Bennis, Warren and Burt Nanus. (1985). *Leaders: The Strategies for Taking Charge.* New York: Harper & Row.

Bennis, Warren. (1989). *On Becoming a Leader.* Reading, MA: Addison-Wesley.

Bentley, Eric. (1957). *A Century of Hero-Worship: A Study of the Idea of Heroism in Carlyle and Nietzsche.* Boston: Beacon Press.

Berger, Jr., Harry. (Spring 1973). "Naive Consciousness and Culture Change: An Essay in Historical Structuralism," *Bulletin of the Midwest Modern Language Association*, 6, No. 1.

Best, Geoffrey. (2001). *Churchill: A Study in Greatness.* London: Hambledon and London.

Blount, Roy, Jr. (2003). *Robert E. Lee.* New York: Viking.

Blumenson, M. (1985). *Patton: The Man Behind the Legend, 1885–1945.* New York: Morrow.

Bolles, Richard N. (1981). *The Three Boxes of Life.* Berkeley, CA: Ten Speed Press.

Boorstin, Daniel J. (1992). *The Creators: A History of Heroes of the Imagination.* New York: Random House.

———. (1987). *The Image: A Guide to Pseudo-Events in America.* New York: Atheneum.

Bowlby, John. (1969). *Attachment.* New York: Basic Books.

———. (1990). *Charles Darwin: A New Life.* New York: Norton.

Brain, R. (1966). *Some Reflections on Genius.* Philadelphia: Lippincott.

Brent, P. (1975). *T. E. Lawrence.* New York: Putnam.

Bridges, T. C. and H. H. Tiltman. (1927). *Heroes of Modern Adventure.* London: Harrap.

Brighton, Terry. (2008). *Patton, Montgomery, Rommel: Masters of War.* New York: Crown.

Bronfenbrenner, U. (1972). *Influences on Human Development.* Hinsdale, IL: Dryden.

Brown, M. (1989). *T. E. Lawrence: The Selected Letters.* New York: Norton.

Brown, M. and Cave, J. (1988). *A Touch of Genius: The Life of T. E. Lawrence*. London: Dent.

Brown, R. (1987). *Analyzing Love*. Cambridge: Cambridge University Press.

Brown, Stuart L. (1994). "Animals at Play." *National Geographic*, 186, No. 6, 2–35.

Buchan, John. (1985) *The Thirty-Nine Steps*. London: Chatton and Windus Ltd.

Bugliosi, Vincent. (1996) *Outrage: The Five Reasons Why O.J. Simpson Got Away with Murder.* New York: Norton.

Byham, William C. (1989). *Zapp! The Lightning of Empowerment*. Pittsburgh, PA: Developmental Dimensions International Press.

Byham, William C. and Jeff Cox. (1994). *HeroZ: Empower Yourself, Your Coworkers, Your Company*. New York: Harmony.

Campbell, Joseph. (1949). *The Hero with a Thousand Faces*. Princeton, NJ: Princeton University Press.

———. (1973). *Myths to Live By*. New York: Viking.

Cancian, R. M. (1987). *Love in America*. Cambridge: Cambridge University Press.

Cannaday, M. (1990). *Bigger than Life: The Creator of Doc Savage*. Bowling Green, OH: Bowling Green State University Press.

Cannadine, David, ed. (1989). *Blood, Toil, Tears and Sweat: The Speeches of Winston Churchill*. Boston: Houghton Mifflin.

———. (2003). *In Churchill's Shadow: Confronting the Past in Modern Britain*. Oxford: Oxford University Press.

Carhart, Tom. (2005). *Lost Triumph: Lee's Real Plan at Gettysburg—and Why it Failed*. New York: Putnam.

Carlyle, Thomas. (1891). *On Heroes, Hero-Worship and The Heroic in History*. London: Chapman and Hall.

Carr, Edward Hallett. (1961). *What is History?* New York: Random House.

Carse, James P. (1986). *Finite and Infinite Games*. New York: Ballantine Books.

———. (1994). *Breakfast at the Victory: The Mysticism of Ordinary Experience*. New York: Harper San Francisco.

Case, R. (1985). *Intellectual Development: Birth to Adulthood*. Orlando, FL: Academic Press.

Cawelti, John G. (1976). *Adventure, Mystery, and Romance: Formula Stories as Art and Popular Culture*. Chicago: The University of Chicago Press.

Churchill, Sarah. (1966). *A Thread in the Tapestry*. New York: Dodd, Mead.

Churchill, Winston S. (1899). *Savrola: A Tale of the Revolution in Laurania*. London: Longmans, Green & Co.

———. (1948). *The Second World War*: 6 vols. New York: Houghton Mifflin.

———. (1965). *A History of the English Speaking Peoples*. New York: Dodd, Mead & Company, Inc.

———. (1990). *Great Contemporaries*. London: Leo Cooper.

———. (1996). *My Early Life: 1874–1904*. New York: Simon and Schuster.

———. (n.d.). *His Wit and Wisdom*, Ed. Jack House. London: Hyperion.

Clarke-Stewart, A., M. Perlmutter and S. Friedman. (1988). *Lifelong Human Development*. New York: Wiley.

Cohen, William A. (1990). *The Art of the Leader*. Englewood Cliffs, NJ: Prentice-Hall.

Conger, Jay A. (1989). *The Charismatic Leader: Behind the Mystique of Exceptional Leadership*. San Francisco: Jossey-Bass.

Coote, Colin and Batchelor, Denzil, eds. (1947). *Maxims and Reflections of the Rt. Hon. Winston S. Churchill, C.H., M.P.* London: Eyre and Spottiswoode, Ltd.

Cornwell. Patricia. (1994). *The Body Farm*. New York: Scribner's.

Covey, Stephen R. (1989). *The Seven Habits of Highly Effective People*. New York: Simon & Schuster.

———. (1991). *Principle-Centered Leadership*. New York: Summit Books.

Cushman, D.P. and R. D. McPhee (1980). *Message-Attitude-Behavior-Relationship*. New York: Academic Press.

Cushman, D.P. and D. D. Cahn. (1985). *Communication in Interpersonal Relationships*. Albany, NY: State University of New York Press.

Deming, W. Edwards. (1986). *Out of Crisis*. Cambridge, MA: Massachusetts Institute of Technology, Center for Advanced Engineering Study.

D'Este, Carlo. (1995). *Patton: A Genius for War*. New York: HarperCollins.

Drucker, P. F. (1967). *The Effective Executive*. New York: Harper & Row.

———. (Jan. 6, 1988). "Leadership: More Doing Than Dash." *Wall Street Journal*.

Dumezil, G. (1970). *The Destiny of the Warrior*. Chicago: The University of Chicago Press.

Dunne, J. S. (1970). *A Search for God in Time and Memory*. London: Collier-Macmillan.

———. (1973). *The City of the Gods*. London: Collier-Macmillan.

Durschmied, Erik. (1999). *The Hinge Factor: How Chance and Stupidity Have Changed History*. New York: Arcade Publishing Inc.

Eliade, Mircea. (1959). *Cosmos and History: The Myth of the Eternal Return*. New York: Harper & Brothers.

Ellroy, James. (1996). *My Dark Places: An L.A. Crime Memoir*. New York: Knopf.

Enright, Dominique, ed. (2002). *The Wit and Wisdom of Jane Austen*. New York: Barnes & Noble.

Erikson, E. H. (1958). *Young Man Luther*. New York: Norton.

———. (1963). *Childhood and Society*. 2nd ed. New York: Norton.

———. (1964). *Insight and Responsibility*. New York: Norton.

———. (1968). *Identity: Youth and Crisis*. New York: Norton.

———. (1968). "On the Nature of Psycho-Historical Evidence: In Search of Gandhi," *Daedalus*, 97.

———. (1969). *Gandhi's Truth*. New York: Norton.

———. (1980). *Identity and the Life Cycle*. New York: Norton.

Erikson, Erik H. and Joan M. Erikson. (1998). *The Life Cycle Completed*. Extended Version. New York: Norton.

Evans, R. I. (1967). *Dialogue with Erik Erikson*. New York: Harper and Row.

Farago, L. (1981). *The Last Days of Patton*. New York: McGraw-Hill.

Fiedler, Leslie. (1960). *Love and Death in the American Novel*. New York: Simon & Schuster.

———. (1982). *What Was Literature?* New York: Simon & Schuster.
Fine, R. (1985). *The Meaning of Love in Human Experience*. New York: Wiley.
Firstbrook, Peter. (2000). *Lost on Everest: The Search for Mallory & Irvine*. London: BBC Worldwide, Ltd.
Fishwick, M. (1969). *The Hero, American Style*. New York: McKay.
Flexner, Stuart Berg, ed. (1993). *Random House Unabridged Dictionary*, 2nd. ed. New York: Random House.
Foucault, Michel. (1965). *Madness and Civilization: A History of Insanity in the Age of Reason*. Trans. Robert Hurley. New York: Random House.
Fraser, Lady Antonia. (1994). Radio interview, The John Dunne Show, October.
Frazer, Sir James. (1890). *The Golden Bough*. New York: Gramercy.
Freeman, G. L. (1973). *Self-fulfillment and Aging*. Watkins Glen, NY: The American Life Foundation & Study Institute.
Freud, S. (1951). *Psychopathology of Everyday Life*. New York: Mentor.
——— (1961). *The Future of an Illusion*. London: Hogarth Press.
Fromm, E. (1956). *The Art of Loving*. New York: Harper & Row.
Frye, Northrop. (1957). *Anatomy of Criticism: Four Essays*. Princeton, NJ: Princeton University Press.
———. (1963). *Fables of Identity: Studies in Poetic Mythology*. New York: Harcourt, Brace and World.
———. (1968). *A Study of English Romanticism*. New York: Random House.
———. (1976). *The Secular Scripture: A Study of the Structure of Romance*. Cambridge, MA: Harvard University Press.
Fuhrman, Mark. (1997). *Murder in Brentwood*. New York: Kensington.
———. (1998). *Murder in Greenwich*. New York: Cliff Street.
———. (2001). *Murder in Spokane*. New York: Cliff Street.
Garnett, David, ed. (1938). *The Letters of T. E. Lawrence*. London: Jonathan Cape.
———. (1951). *The Essential T. E. Lawrence*. London: Jonathan Cape.
Geertz, Clifford. (1973). *The Interpretation of Cultures*. New York: Basic Books.
Gesell, Arnold. (1948). *Studies in Child Development*. New York: Harper & Brothers.
———. (1956). *Youth: the Years from Ten to Sixteen*. New York: Harper.
———. (1972). *Infant Development: the Embryology of Early Human Behavior*. Westport, CT: Greenwood Press.
Gibbon, Peter H. (2002). *A Call to Heroism: Renewing America's Vision of Greatness*. New York: Atlantic Monthly Press.
Gilbert, Sandra and Susan Gubar. (1979). *The Madwoman in the Attic: The Woman Writer and the Nineteenth-Century Literary Imagination*. New Haven, CT: Yale University Press.
Gleick, James. (1992). *Genius: The Life and Science of Richard Feynman*. New York: Pantheon.
Goodrich, Norma Lorre. (1991). *Guinevere*. New York: Harper Collins.
———. (1992). *The Holy Grail*. New York: Harper Collins.
———. (1993). *Heroines*. New York: Harper Collins.
Goodwin, Doris Kearns. (1994). *No Ordinary Time—Franklin and Eleanor Roosevelt: The Home Front in World War II*. New York: Simon & Schuster.

———. (2005). *Team of Rivals: The Political Genius of Abraham Lincoln*. New York: Simon & Schuster.
Graebner, Walter. (1965). *My dear Mister Churchill*. London: Michael Joseph, Ltd.
Grange, Kathleen M. (1962). "Samuel Johnson's Account of Certain Psychoanalytic Concepts," *Journal of Nervous and Mental Disease*, CXXXV, 93–98.
Graves, Richard Perceval. (1976). *Lawrence of Arabia and his World*. New York: Scribner's.
Graves, Robert and Liddell Hart. (1963). *T. E. Lawrence to his Biographers*. Garden City, NY: Doubleday.
Greene, John C. (1963). *Darwin and the Modern World View*. New York: New American Library.
———. (1981). *Science, Ideology and World View: Essays in the History of Evolutionary Ideas*. Berkeley and Los Angeles: University of California Press.
Gross, Gloria Sybil. (2002). *In a Fast Coach with a Pretty Woman: Jane Austen and Samuel Johnson*. New York: AMS Press.
Grunwald, B. B., & H. V. McAbee. (1985). *Guiding the Family: Practical Counseling Techniques*. Muncie, IN: Accelerated Development.
Guedalla, Philip. (1941). *Mr. Churchill, A Portrait*. London: Hodder and Stoughton Limited.
Hadas, Moses and Smith, Morton. (1965), *Heroes and Gods: Spiritual Biographies in Antiquity*. New York: Harper and Row.
Haddon, Mark. (2003). *The Curious Incident of the Dog in the Night-Time*. New York: Doubleday.
Haffner, Sebastian. (2003). *Churchill*. London: Haus Publishing Limited.
Halberstam, David (1969). *The Best and the Brightest*. New York: Random House.
Halperin, John. (1984). *The Life of Jane Austen*. Baltimore, MD: Johns Hopkins University Press.
Hardison, O. B., Jr. (1989). *Disappearing Through the Skylight: Culture and Technology in the Twentieth Century*. New York: Viking.
Harris, Thomas. (1975). *Black Sunday*. New York: G.P. Putnam's Sons.
———. (1981). *Red Dragon*. New York: Putnam.
———. (1988). *The Silence of the Lambs*. New York: St. Martin's Press.
———. (1999). *Hannibal*. New York: Delacorte.
———. (2006). *Hannibal Rising*. New York: Delacorte.
Harris, William H. and Judith S. Levey, eds. (1975). *The New Columbia Encyclopedia*. New York: Columbia University Press.
Hart, L. (1934). *'T. E. Lawrence' in Arabia and After*. London: Jonathan Cape.
———. (1934). *The Man Behind the Legend: Colonel Lawrence (of Arabia)*. New York: Halcyon.
Hayward, Justin and Ray Thomas. (1991). "Never Blame the Rainbows for the Rain," *Keys of the Kingdom*. London: Bright Music Ltd.
Heath, F. W., ed. (1966). *Churchill in His Own Words: Years of Adventure*. New York: Capricorn.
Herndon, Booton. (1967). *The Unlikeliest Hero*. Mountain View, CA: Pacific Press.

Hersey, P. and Stinson, J. (1980). *Perspectives in Leader Effectiveness*. Athens, OH: Ohio University Center for Leadership Studies.

Hinde, R. A. (1979). *Towards Understanding Relationships*. London: Academic Press.

———. (1987). *Individuals, Relationships & Culture: Links Between Ethology and the Social Sciences*. Cambridge: Cambridge University Press.

Hiney, Tom. (1997). *Raymond Chandler: A Biography*. New York: Atlantic Monthly Press.

Hodges, Andrew. (1983). *Alan Turing: The Enigma*. New York: Simon & Schuster.

Honan, Park. (1987). *Jane Austen: Her Life*. New York: Fawcett Columbine.

Hook, S. (1955). *The Hero in History: A Study in Limitation and Possibility*. Boston: Beacon Press.

Hughes, F. P. and Noppe, L. D. (1985). *Human Development Across the Life Span*. St. Paul, MN: West.

Huizinga, Johan. (1955). *Homo Ludens: A Study of the Play-Element in Culture*. Boston: Beacon Press.

Humes, James C., ed. (1994). *The Wit and Wisdom of Winston Churchill*. New York: HarperCollins.

Huntford, Roland. (1985). *Shackleton*. New York: Carroll & Graf.

Hyde, H. M. (1978). *Solitary in the Ranks: Lawrence of Arabia as Airman and Private Soldier*. New York: Atheneum.

Jacoby, Russell. (1987). *The Last Intellectuals: American Culture in the Age of Academe*. New York: Basic Books.

———. (1994). *Dogmatic Wisdom: How the Culture Wars Divert Education and Distract America*. New York: Doubleday.

Jaffe, Dennis T. and Scott, Cynthia D. (1984). *From Burnout to Balance: A Workbook for Personal Self-Renewal*. New York: McGraw-Hill.

James, L. (1990). *The Golden Warrior: The Life and Legend of Lawrence of Arabia*. London: Weidenfeld and Nicolson.

James, William (1899). *Talks to Teachers on Psychology: and to Students on Some of Life's Ideals*. New York: H. Holt and Co.

Jameson, Frederic. (1991). *Postmodernism, or, the Cultural Logic of Late Capitalism*. Durham, NC: Duke University Press.

Jenkins, Roy. (2001). *Churchill: A Biography*. New York: Farrar, Straus and Giroux.

———. (2002). *Why Churchill's Speech Was A Shot Which Rang Around The World*. Fulton, MO: The Winston Churchill Memorial and Library.

Johnson, Paul. (1988). *Intellectuals*. New York: Harper & Row.

———. (2007). *Heroes*. New York: HarperCollins.

———. (2009). *Churchill*. New York: Viking.

Johnson, Samuel. (1773). *A Dictionary of the English Language*, 4th ed. London: W. Strahan.

———. (1958). *Diaries, Prayers, and Annals*. Ed. E. L. McAdam, Jr. with Donald and Mary Hyde. New Haven: Yale University Press.

———. (1968). *Johnson on Shakespeare*. Ed. Arthur Sherbo, 2 vols. New Haven: Yale University Press.

———. (1990). *Rasselas and Other Tales*. Ed. Gwin J. Kolb. New Haven: Yale University Press.
Jones, Max, ed. (2006). *Robert Falcon Scott Journals: Captain Scott's Last Expedition*. Oxford: Oxford University Press.
Jones, Richard Foster. (1961). *Ancients and Moderns: A Study of the Rise of the Scientific Movement in Seventeenth-Century England*. 2nd ed. St. Louis: Washington University Press.
Jost, David A., project director. (1993). *The American Heritage College Dictionary*, 3rd. ed. Boston: Houghton Mifflin.
Jung, C.G. (1933). *Modern Man in Search of a Soul*. London: K. Paul, Trench, Trubner, & Co., Ltd.
———. (1958). *The Undiscovered Self*. Trans. R.F.C. Hull. Boston: Little, Brown.
———. (1959). *The Archetypes and the Collective Unconscious*. Trans. R.F.C. Hull. New York: Pantheon.
———. (1969). *Psychology and Education*. Trans. R.F.C. Hull. Princeton, NJ: Princeton University Press.
———. (1970). *The Development of Personality*. Trans. R.F.C. Hull. Princeton, NJ: Princeton University Press.
———. (1971). *Four Archetypes: Mother, Rebirth, Spirit, Trickster*. Trans. R.F.C. Hull. Princeton, NJ: Princeton University Press.
Keegan, John. (1976). *The Face of Battle*. London: Jonathan Cape.
———. (1988). *The Mask of Command*. New York: Penguin.
———. (2002). *Winston Churchill*. New York: Viking.
———. (2009). *The American Civil War: A Military History*. New York: Knopf.
Kegan, R. (1982). *The Evolving Self: Problem and Process in Human Development*. Cambridge, MA: Harvard University Press.
Kernan, Alvin. (1990). *The Death of Literature*. New Haven, CT: Yale University Press.
Kelling, G. W. (1979). *Blind Mazes*. Chicago, IL: Nelson-Hall.
Kelly, C. Brian and Ingrid Smyer. (2008). *Best Little Stories from the Life and Times of Winston Churchill*. Nashville, TN: Cumberland House Publishing, Inc.
King, Dean. (2000). *Patrick O'Brian: A Life Revealed*. New York: Henry Holt & Co.
Klingaman, William K. (1988). *1941: Our Lives in a World on the Edge*. New York: Harper & Row.
Knight, R. D. (1985). *Colonel T. E. Lawrence (Lawrence of Arabia) Visits Mr. & Mrs. Thomas Hardy*. Weymouth: Bat & Ball Press.
Knightley, P. and C. Simpson. (1969). *The Secret Lives of Lawrence of Arabia*. London: Nelson.
Knightley, P. (1990). "Desert Warriors," *M Inc*.
Kobal, J. (1973). *Romance and the Cinema*. London: Studio Vista.
Kohlberg, L. (1969). "Stage and Sequence: The Cognitive-Developmental Approach to Socialization," in *Handbook of Socialization Theory and Research*. Ed. D. A. Goslin. Chicago: Rand McNally.
Kossen, S. (1983). *The Human Side of Organizations*. New York: Harper & Row.

Kotre, John N. (1984). *Outliving the Self: Generativity and the Interpretation of Lives*. Baltimore: Johns Hopkins University Press.

Kotre, John N. and Elizabeth Hall. (1990). *Seasons of Life: Our Dramatic Journey from Birth to Death*. Boston, MA: Little, Brown.

Kouzes, James M. and Barry Posner. (1987). *The Leadership Challenge: How to Get Extraordinary Things Done in Organizations*. San Francisco: Jossey-Bass.

Krakauer, Jon. (1992). *Eiger Dreams: Ventures Among Men and Mountains*. Guilford CT: Lyons Press.

———. (1996). *Into The Wild*. New York: Villard.

———. (1997). *Into Thin Air*. New York: Villard.

Kuhn, Thomas. (1970). *The Structure of Scientific Revolutions*. 2nd ed. Chicago: The University of Chicago Press.

Lawrence, A. W., ed. (1937). *T. E. Lawrence by His Friends*. London: Jonathan Cape.

———. ed. (1962). *Letters to T. E. Lawrence*. London: Jonathan Cape.

Lawrence, T. E. (1927). *Revolt in the Desert*. London: Jonathan Cape.

———. (1935). *Seven Pillars of Wisdom: A Triumph*. London: Jonathan Cape.

———. (1955). *T. E. Lawrence The Mint: A day-book of the R.A.F. Depot between August and December 1922 with later notes by 352087 A/c ROSS.*

———. (1971). *Minorities*. Ed. Jeremy M. Wilson. London: Jonathan Cape.

———. (1986. Orig. published 1936). *Crusader Castles*. London: Michael Haag.

Lawrence, T.E. and C. Leonard Woolley. (1936). *The Wilderness of Zin*. New York: Scribner's.

Lerner, R. M. (1986). *Concepts and Theories of Human Development*. New York: Random House.

Lepp, I. (1963). *The Psychology of Loving*. Baltimore: Helicon.

Lethem, Jonathan. (1999). *Motherless Brooklyn*. New York: Doubleday.

Levine, Arthur. (1980). *When Dreams and Heroes Died*. San Francisco: Jossey-Bass.

Lewis, C. S. (1960). *The Four Loves*. New York: Harcourt, Brace and World.

Lewis, Hunter (2000). *A Question of Values: Six Ways We Make the Personal Choices That Shape Our Lives*. Mount Jackson, VA: Axios Press.

———. (2000). *The Beguiling Serpent: A Re-evaluation of Emotions And Values*. Mount Jackson, VA: Axios Press.

Lewis, Jon E., ed. (2002). *The Mammoth Book of Heroes: True Tales of Courage from Ancient Times to September 11, 2001*. New York: Carroll & Graf.

Lewis, M. and Rosenblum, L. A. (1978). *The Development of Affect*. New York: Plenum Press.

Lewis, R. W. B. (1955). *The American Adam*. Chicago: The University of Chicago Press.

Liddy, G. Gordon. (1980). *Will: The Autobiography of G. Gordon Liddy*. New York: St. Martin's Press.

Lodge, John. (1972). "Isn't Life Strange," *Seventh Sojourn*. London: Threshold Music Ltd.

Lorenz, K. Z. (1966). *On Aggression*. New York: Harcourt, Brace and World.
Lovejoy, Arthur O. (1936). *The Great Chain of Being: A Study of the History of an Idea*. Cambridge: Harvard University Press.
Lovell, Mary S. (1989). *The Sound of Wings: The Life of Amelia Earhart*. New York: St. Martin's.
Lubin, H. (1968). *Heroes and Anti-Heroes*. San Francisco: Chandler.
Lucas, F. L. (1951). *Literature and Psychology*. London: Cassell and Company.
Lukacs, John. (1999). *Five Days in London: May 1940*. New Haven: Yale University Press
———. (2001). *The Duel: The Eighty-Day Struggle Between Churchill and Hitler*. New Haven: Yale University Press.
———. (2002). *Churchill: Visionary. Statesman. Historian*. New Haven: Yale University Press.
Maccoby, E. E. (1980). *Social Development: Psychological Growth and the Parent-Child Relationship*. New York: Harcourt Brace.
Macdonald, Dwight. (1961). *Masscult & Midcult*. New York: Partisan Review.
Mack, J. E. (1976). *A Prince of Our Disorder: The Life of T. E. Lawrence*. Boston, MA: Little, Brown.
MacLean, Norman (1976). *A River Runs Through It*. Chicago: University of Chicago Press.
McConnell, Frank. (1979). *Storytelling and Mythmaking*. New York: Oxford University Press.
McCullough, David. (1992). *Brave Companions: Portraits in History*. New York: Prentice Hall.
McLuhan, Marshall. (1962). *The Gutenberg Galaxy: The Making of Typographic Man*. Toronto: University of Toronto Press.
McLuhan, Marshall and Quentin Fiore. (1967). *The Medium is the Massage*. New York: Random House.
McLuhan, Marshall and Bruce R. Powers. (1989). *The Global Village: Transformations in World Life and Media in the 21st Century*. New York: Oxford University Press.
McNeil, E. B. (1971). *Readings in Human Socialization*. Belmont, CA: Brooks/Cole.
Mahoney, Richard J. and Shera Dalin, eds. (2005). *The Quotable Winston Churchill*. Fulton, MO: Winston Churchill Memorial and Library.
Manchester, William. (1978). *American Caesar*. Boston, MA: Little, Brown.
———. (1983). *The Last Lion, Winston Spencer Churchill: Visions of Glory, 1874–1932*. Boston, MA: Little, Brown.
———. (1988). *The Last Lion, Winston Spencer Churchill: Alone, 1932–1940*. Boston, MA: Little, Brown.
Maraniss, David. (2008). *Rome 1960: The Olympics That Changed the World*. New York: Simon and Schuster.
Marfo, Kofi, ed. (1988). *Parent-Child Interaction and Developmental* Disabilities: Theory, Research and Intervention. New York: Praeger.
May, Rollo. (1953). *Man's Search for Himself*. New York: Norton.

———. (1969). *Love and Will*. New York: Norton.
———. 1975). *The Courage to Create*. New York: Norton.
———. (1991). *The Cry for Myth*. New York: Norton.
Maynard, Joyce. (1998). *At Home in the World: A Memoir*. New York: Picador.
Megill, Allan. (1985). *Prophets of Extremity: Nietzsche, Heidegger, Foucault, Derrida*. Berkeley and Los Angeles: University of California Press.
Meyers, Jeffrey. (1973). *The Wounded Spirit: A Study of SEVEN PILLARS OF WISDOM*. London: Martin Brian & O'Keeffe.
Miller, Patricia H. (1989). *Theories of Developmental Psychology*. 2nd ed. New York: Freeman and Company.
Milton, C.R. (1981). *Human Behavior in Organizations: Three Levels of Behavior*. Englewood Cliffs, NJ: Prentice-Hall.
Mish, Frederick C., ed. (1993). *Merriam-Webster's Collegiate Dictionary*, 10th ed. Springfield, MA: Merriam-Webster.
Moir, Anne and David Jessel. (1991). *Brain Sex: The Real Difference Between Men and Women*. New York: Carol Publishing Group.
Monk, Ray. (1990). *Ludwig Wittgenstein: The Duty of Genius*. New York: Free Press.
Moran, Charles Wilson, Lord. (1966). *Churchill: Taken from the Diaries of Lord Moran, The Struggle for Survival, 1940–1965*. Boston: Houghton Mifflin.
Morrell, Margot and Stephanie Capparell. (2001). *Shackleton's Way*. New York: Viking.
Mousa, Suleiman. (1966). *T. E. Lawrence: An Arab View*. London: Oxford University Press.
Muller, James W., ed. (1997). *Churchill As Peacemaker* (Woodrow Wilson Center Series). Cambridge: Cambridge University Press.
Murdock, Maureen. (1990). *The Heroine's Journey*. Boston: Shambhala.
Myer, Valerie Grosvenor. (1997). *Jane Austen, Obstinate Heart: A Biography*. New York: Arcade.
Nel, Elizabeth. (2007). *Winston Churchill by his Personal Secretary*. Lincoln, NE: iUniverse, Inc.
Nicolson, Nigel. (1991). *The World of Jane Austen*. London: Weidenfied and Nicolson.
Nietzsche, Friedrich. (1989). *Beyond Good and Evil*. Trans. Walter Kaufmann. New York: Vintage.
Nokes, David. (1997). *Jane Austen: A LIfe*. New York: Farrar, Straus and Giroux.
Nozick, Robert. (1989). *The Examined Life: Philosophical Meditations*. New York: Simon & Schuster.
Nye, Roger H. (1993). *The Patton Mind*. Garden City Park, NY: Avery.
Ocampo, V. (1963). *338171 T. E. Lawrence of Arabia*. London: Gollancz.
O'Donnell, Thomas J. (1979). *The Confessions of T. E. Lawrence: The Romantic Hero's Presentation of Self*. Athens, OH: Ohio University Press.
Offer, Daniel and Melvin Sabshin, eds. (1984). *Normality and the Life Cycle: A Critical Integration*. New York: Basic Books.
Olson, Lynne. (2007). *Troublesome Young Men*. New York: Farrar, Straus and Giroux.

Paglia, Camille. (1990). *Sexual Personae: Art and Decadence from Nefertiti to Emily Dickinson*. New Haven, CT: Yale University Press.
———. (1992). *Sex, Art, and American Culture: Essays*. New York: Vintage.
Pascarella, Perry and Mark A. Frohman. (1989). *The Purpose-Driven Organization: Unleashing the Power of Direction and Commitment*. San Francisco: Jossey-Bass.
Patton, George S., Jr. (1947). *War As I Knew It*. Boston, MA: Houghton Mifflin.
Patton, Robert H. (1994). *The Pattons: A Personal History of an American Family*. New York: Crown Publishers, Inc.
Pearson, C. S. (1986). *The Hero Within: Six Archetypes We Live By*. New York: Harper & Row.
Peters, Tom. (1988). *Thriving on Chaos: Handbook for Management Revolution*. New York: Alfred A. Knopf.
Petrullo, L. and Bass, B. M. (1961). *Leadership and Interpersonal Behavior*. New York: Holt, Rinehart.
Piaget, Jean. (1923). *The Language and Thought of the Child*. New York: Harcourt, Brace.
———. (1947). *The Psychology of Intelligence*. New York: Harcourt, Brace.
———. (1954). *The Construction of Reality in the Child*. New York: Basic Books.
———. (1997). *How the Mind Works*. New York: Norton.
Piaget, Jean and Barbel Inhelder. (1969). *The Psychology of the Child*. New York: Basic Books.
Pinker, Steven. (1994). *The Language Instinct: How the Mind Creates Language*. NewYork: Morrow.
———. (2002). *The Blank Slate: The Modern Denial of Human Nature*. New York: Viking.
———. (2007). *The Stuff of Thought: Language as a Window into Human Nature*. New York: Viking.
Pinkola-Estés, C. (1992). *Women Who Run With the Wolves*. New York: Ballantine.
Plato. (1960). *Gorgias*. Trans. Walter Hamilton. Harmondsworth: Penguin.
———. (1983) *Meno*. Trans. Benjamin Jowett, intro. Fulton H. Anderson. New York: Liberal Arts Press.
———. (2002). *Phaedrus*. Trans. Robin Waterfield. New York: Oxford University Press.
Popkin, Richard H. (1964). *The History of Scepticism from Erasmus to Descartes*. rev. ed. New York: Humanities Press.
Province, Charles M. (1995). *Patton's One-Minute Messages: Tactical Leadership Skills for Business Management*. Novato, CA: Presidio Press.
Puryear, Jr., Edgar F. (1971). *Nineteen Stars: A Study in Military Character and Leadership*. Orange, VA: Green Publishers.
Raglan, L. (1975). *The Hero: A Study in Tradition, Myth and Drama*. New York: Vintage.
Rank, O. (1932). *The Myth of the Birth of the Hero and Other Writings*. Ed. P. Freund. New York: Vintage.
Rattigan, T. (1960). *Ross: A Dramatic Portrait*. London: Hamish Hamilton.

Ressler, Robert K. and Tom Shachtman. (1992) *Whoever Fights Monsters*. New York: St. Martin's Press.

———. (1997). *I Have Lived in the Monster: A Report from the Abyss*. New York: St. Martin's Press.

Rigney, Barbara Hill. (1978). *Madness and Sexual Politics in the Feminist Novel*. Madison, WI: The University of Wisconsin Press.

Roberts, Andrew. (2003). *Hitler and Churchill*. London: Weidenfeld & Nicolson.

Roberts, W. (1993). *Leadership Secrets of Attila the Hun*. New York: Warner.

Robinson, Daniel N. (1979). *Systems of Modern Psychology: A Critical Sketch*. New York: Columbia University Press.

Rorty, Richard. (1979). *Philosophy and the Mirror of Nature*. Princeton, NJ: Princeton University Press.

Rose, P. (1979). *Socialization and the Life Cycle*. New York: St. Martin's Press.

Rubin, Gretchen. (2003). *Forty Ways to Look at Winston Churchill: A Brief Account of a Long Life*. New York: Ballantine.

Salkind, N. J. (1981). *Theories of Human Development*. 2nd ed. New York: Wiley.

Salusinszky, Imre. (1987). *Criticism in Society*. New York: Methuen.

Sandys, Celia. (1994). *From Winston With Love and Kisses: The Young Churchill*. London: Sinclair-Stevenson.

———. (2000). *Churchill: Wanted Dead or Alive*. New York: Carroll & Graf Publishers.

———. (2003). *Chasing Churchill: The Travels of Winston Churchill*. New York: Carroll & Graf Publishers.

Sandys, Celia and Jonathan Littman. (2003). *We Shall Not Fail: The Inspiring Leadership of Winston Churchill*. New York: Portfolio.

Sayer, George. (1988). *Jack: C. S. Lewis and His Times*. New York: Harper & Row.

Sergiovanni, T. J. and J. E. Corbally. (1984). *Leadership and Organizational Culture*. Urbana, IL: University of Illinois Press.

Shaffer, D. R. (1979). *Social and Personality Development*. Monterey, CA: Brooks/Cole.

Shaara, Jeff. (1996). *Gods and Generals*. New York: Ballantine Books.

———. (1998). *The Last Full Measure*. New York: Ballantine Books.

Shaara, Michael. (1974). *The Killer Angels: A Novel*. New York: David McKay.

Shackleton, Ernest. (1998). *South: A Memoir of the Endurance Voyage*. New York: Carroll & Graf.

Shaw, G. B. (1930). *Man and Superman*. New York: William H. Wise & Co.

———. (1934). *Too True To Be Good, Village Wooing & On the Rocks; Three Plays by Bernard Shaw*. London: Constable and Company, Ltd.

Shaw, T. E. (T. E. Lawrence). (1932). *The Odyssey of Homer*. New York: Oxford University Press.

Sheehy, Gail. (1976). *Passages: Predictable Crises of Adult Life*. New York: Dutton.

Shenk, Joshua Wolf. (2005). *Lincoln's Melancholy: How Depression Challenged a President and Fueled His Greatness*. New York: Houghton Mifflin.

Sherman, Robert and Don Dinkmeyer, et al. (1987). *Systems of Family Therapy: An Adlerian Integration*. New York: Bruner/Mazel.

Simpson, George Gaylord. (1950). *The Meaning of Evolution: A Study of the History of Life and of Its Significance for Man*. New Haven, CT: Yale University Press.

Singer, I. (1987). *The Nature of Love: The Modern World*. Chicago: The University of Chicago Press.

Skinner, B. F. (1948). *Walden Two*. New York: Macmillan.

———. (1971). *Beyond Freedom and Dignity*. New York: Knopf.

Smith, Henry Nash. (1950). *Virgin Land*. Cambridge, MA: Harvard University Press.

Soames, Mary. (1990). *Winston Churchill: His Life As A Painter*. Boston: Houghton Mifflin.

———. ed. (1998). *Winston and Clementine: The Personal Letters of the Churchills*. Boston: Houghton Mifflin.

———. (2003) *Clementine Churchill: The Biography of a Marriage*. New York: Mariner Books.

———. (2004). *Churchill and the Great Republic*. London: D. Giles Ltd.

Solomon, Robert C. (1988). *Continental Philosophy since 1750: The Rise and Fall of the Self*. Oxford: Oxford University Press.

Solomon, Susan. (2001). *The Coldest March: Scott's Fatal Antarctic Expedition*. New Haven: Yale University Press.

Solzhenitsyn, Aleksandr. (1993). "How the Cult of Novelty Wrecked the 20th Century," *American Arts Quarterly*, 10.

Spenser, Edmund. (1977). *The Faerie Queene*, ed. A. C. Hamilton. London: Longman.

Spufford, Francis. (1997). *I May Be Some Time: Ice and the English Imagination*. New York: Picador.

Steiner, George. (1986). *Real Presences*. Cambridge: Cambridge University Press.

Sternberg, R. J. and M. L. Barnes. (1988). *The Psychology of Love*. New Haven, CT: Yale University Press.

Stevens, Anthony. (1983). *Archetypes: A Natural History of the Self*. New York: Quill.

Stiller, Richard. (1997). *Influence As Power*. Santa Clara: Sun Microsystems.

Stillman, Richard J. (1997). *General Patton's Timeless Leadership Principles: Your Practical Guide For a Successful Career and Life*. New Orleans: Richard J. Stillman Co.

Stone, L. (1977). *The Family, Sex, and Marriage in England, 1500–1800*. New York: Harper & Row.

Storr, A. (1988). *Churchill's Black Dog, Kafka's Mice, and Other Phenomena of the Human Mind*. New York: Grove Press.

Strachey, Lytton. (1921). *Queen Victoria*. London: Chatto and Windus.

———. (1928). *Elizabeth and Essex: A Tragic History*. London: Chatto and Windus.

———. (1933) *Eminent Victorians*. New York: The Modern Library.

Strawson, John. (1998). *Churchill and Hitler: In Victory and Defeat*. New York: Fromm International.

Stuart, G. (1938). *The Achievement of Personality*. New York: Macmillan.

Sturridge, Charles, dir. (2002) *Kenneth Branagh: Shackleton*. A&E dvd set, with the original production, the Biography Channel's *Shackleton*, and other subsidiary material.

Sullenberger, Captain Chesley "Sully" with Jeffrey Zaslow. (2009). *Highest Duty: My Search for What Really Matters*. New York: William Morrow.

Tabachnick, S. E., ed. (1984). *The T. E. Lawrence Puzzle*. Athens, GA: University of Georgia Press.

Taylor, Charles. (1989). *Sources of the Self: The Making of the Modern Identity*. Cambridge, MA: Harvard University Press.

Taylor, Robert Lewis. (1952). *Winston Churchill: An Informal Study of Greatness*. Garden City, NY: Doubleday.

Teilhard de Chardin, Pierre. (1959). *The Phenomenon of Man*. New York: Harper & Brothers.

Thomas, Lowell. (1924). *With Lawrence in Arabia*. New York: P. F. Collier & Son.

———. (1927). *The Boys' Life of Colonel Lawrence*. New York: Century.

Thoreau, Henry David. (1961). *Walden or Life in the Woods*. Ed. Perry Miller. New York: Signet.

Tichy, Noel M. and Mary Anne Devanna. (1986). *The Transformational Leader*. New York: John Wiley & Sons.

Toffler, Alvin. (1970). *Future Shock*. New York: Random House.

Tomalin, Claire. (1997). *Jane Austen: A Life*. New York: Knopf.

Tompkins, Jane. (1992). *West of Everything: The Inner Life of Westerns*. New York: Oxford University Press.

Tornstam, Lars. (2005). *Gerotranscendence: A Developmental Theory of Positive Aging*. New York: Springer Publ. Co.

Toulmin, Stephen. (1990). *Cosmopolis: The Hidden Agenda of Modernity*. New York: Free Press.

Trulock, Alice Rains. (1992). *In the Hands of Providence*. Chapel Hill, NC: The University of North Carolina Press.

Tucker, Robert C. (1968). "The Theory of Charismatic Leadership," *Daedalus*, 97.

Turner, Frederick Jackson. (1920). *The Frontier in American History*. New York: H. Holt and Co.

Vander Zanden, J. W. (1989). *Human Development*. New York: Knopf.

Villars, Jean Beraud. (1958). *T. E. Lawrence or The Search for the Absolute*. London Sidgwick and Jackson.

Wallach, Janet. (1996). *Desert Queen*. London: Weidenfeld & Nicolson.

Walster, E. and G. W. Walster. (1978). *A New Look at Love*. Reading, MA: Addison-Wesley.

Walton, Mary. (1986). *The Deming Management Method*. New York: Putnam.

———. (1991). *Deming Management at Work*. New York: Putnam.

Ward, A. C. (1943). *A Literary Journey Through Wartime Britain*. New York: Oxford University Press.

Wecter, D. (1938). *The Hero in America*. Ann Arbor, MI: The University of Michigan Press.

Weintraub, Stanley and Rodelle, eds. (1967). *Evolution of a Revolt: Early Postwar Writings of T. E. Lawrence*. University Park, PA: The Pennsylvania State University Press.

Weintraub, Stanley. (1963). *Private Shaw and Public Shaw: A Dual Portrait of Lawrence of Arabia and G. B. S.*, London: Jonathan Cape.

Wert, Jeffrey D. (1993). *General James Longstreet: The Confederacy's Most Controversial Soldier—A Biography*. New York: Simon & Schuster.

Westfall, Richard S. (1980). "Isaac Newton in Cambridge: The Restoration University and Scientific Creativity" in *Culture and Politics From Puritanism to the Enlightenment*, ed. Perez Zagorin. Berkeley and Los Angeles: University of California Press.

Weston, J. L. (1957). *From Ritual to Romance*. Garden City, NY: Doubleday.

Wilde, Oscar. (1894). *Intentions*. New York: Dodd, Mead.

Williamson, H. (1941). *Genius of Friendship: 'T. E. Lawrence'*. London: Faber and Faber.

Wills, Garry. (1992). *Lincoln at Gettysburg*. New York: Simon & Schuster.

———. (1994). *Certain Trumpets*. New York: Simon & Schuster.

Wilshire, Bruce. (1990). *The Moral Collapse of the University: Professionalism, Purity, and Alienation*. Albany: State University of New York Press.

Wilson, Charles, 1st Baron Moran. (1966). *Churchill: Taken from the Diaries of Lord Moran, The Struggle for Survival, 1940–1965*. Boston: Houghton Mifflin.

Wilson, Edward O. (1975). *Sociobiology: The New Synthesis*. Cambridge, MA: Harvard University Press.

———. (1978). *On Human Nature*. Cambridge, MA: Harvard University Press.

———. (1991). "Biological and Human Determinants of the Survival of Species," in *Individuality and Cooperative Action*, ed. Joseph E. Earley. Washington, D.C.: Georgetown University Press.

Wilson, J. (1989). *Lawrence of Arabia: The Authorised Biography of T. E. Lawrence*. London: Heinemann.

Winik, Jay (2001). *April 1865: The Month That Saved America*. New York: HarperCollins.

Worsley, F.A. (1999). *Shackleton's Boat Journey*. London: Pimlico.

Yardley, M. (1989). *Backing into the Limelight: A Biography of T. E. Lawrence*. London: Harrap.

Yeats, William Butler. (1965). *The Autobiography of William Butler Yeats*. New York: Collier.

Zweig, P. (1974). *The Adventurer*. Princeton, NJ: Princeton University Press.

Index

Breinigsville, PA USA
16 September 2010

245571BV00002B/2/P